P9-DVY-695

The particle hunters

The particle hunters

YUVAL NE'EMAN
Sackler Faculty of Exact Sciences,
Wolfson Distinguished Chair in Theoretical Physics,
Tel Aviv University, and Center for
Particle Theory, University of Texas, Austin

YORAM KIRSH
Physics Group, Everyman's University, Tel Aviv

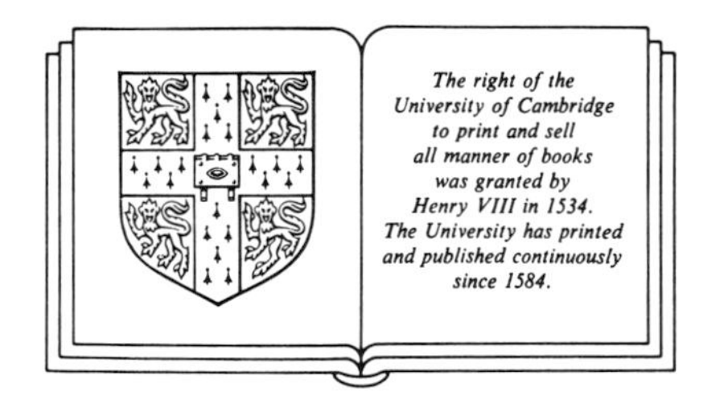

CAMBRIDGE UNIVERSITY PRESS

CAMBRIDGE
NEW YORK NEW ROCHELLE
MELBOURNE SYDNEY

Published by the Press Syndicate of the University of Cambridge
The Pitt Building, Trumpington Street, Cambridge CB2 1RP
32 East 57th Street, New York, NY 10022, USA
10 Stamford Road, Oakleigh, Melbourne 3166, Australia

Original Hebrew version published by Massada,

This updated version first published in English by Cambridge University Press 1986

Reprinted 1987

Printed in Great Britain at the University Press, Cambridge

British Library cataloguing in publication data

Ne'eman, Yuval
The particle hunters
1. Particles (Nuclear physics)
I. Title II. Kirsh, Yoram
539.7'21 QC793.2

Library of Congress cataloguing in publication data

Ne'eman, Yuval.
The particle hunters.

Translation of: Tsayade ha-helkikim.
Includes index.
1. Nuclear physics—Popular works. I. Kirsh, Y.
II. Title.
QC778.N3813 1986 539.7'21 86-21265

ISBN 0 521 30194 7 hard covers
ISBN 0 521 31780 0 paperback

PN

Dedicated to our wives, Dvora and Malka

Contents

Preface

Since its very beginning, physics has pondered the question of what are the fundamental building blocks of matter. In the last century all matter was shown to be composed of atoms. In this century, science has taken a further step forward, revealing the internal structure of the atom. Today, the fact that the atomic nucleus consists of protons and neutrons, and that electrons move around it, is common knowledge. But the story does not end there. Over the past 40 or 50 years it has been found that in collisions between sub-atomic particles additional particles are formed, whose properties and behaviour can shed light on the basic laws of physics. All these particles were called – somewhat unjustifiably – elementary particles.

The study of the sub-atomic structure has led to the development of two branches of modern physics: nuclear physics, which is concerned with furthering the understanding of the atomic nucleus as a whole and the processes going on inside it, and elementary particle physics, the new frontier, which deals with the properties and structure of the various particles themselves and the interactions between them. Up until the end of the 1940s, elementary particle physics had been regarded as a part of nuclear research, but in the 1950s it achieved the status of an independent branch of physics.

The discoveries in the field of nuclear physics have had far-reaching effects on our lives, and everyone is familiar with their applications: on the one hand, nuclear weapons of mighty destructive force, and on the other, nuclear power stations which might solve the energy crisis, and artificial radioactive isotopes employed in medicine, agriculture and industry. In short, nuclear physics is well known to the public at large. But the same is not true of elementary particle physics. Every so often the discovery of a new particle makes the headlines, but it is difficult for anyone outside the profession to follow developments and appreciate their importance. Part of the blame lies in the fact that although particle research is today regarded as the frontier of physics, there are few up-to-date comprehensive texts on the subject for the non-professional. The present book sets out to help fill that gap.

It is our pleasant duty to express our thanks to Dr Simon Capelin of Cambridge University Press and to Professor Alan D. Martin for many valuable suggestions that contributed greatly to the book. We would also like to thank Jenni Coombs for her help in editing the book, Jacob Haimowitz who has read the typescript and made important remarks and Esther Berger for her help in translating a part of the typescript from Hebrew.

YUVAL NE'EMAN AND YORAM KIRSH *Tel Aviv, January, 1986*

Chapter One

The building blocks of the atom

1.1 The beginnings of atomic research

The word 'atom' is derived from the Greek '*atomos*', meaning indivisible. In about the year 400 BC the Greek philosopher Democritus postulated that all matter was made up of minute particles which could not be destroyed or broken up. He was unable to perform any experiment to support his hypothesis, but this concept could account for the fact that different substances had different densities: the more the atoms were compressed the denser and heavier the substance became. A few Greek sages accepted the atomic theory of Democritus, but the great majority adopted the view of Aristotle, who believed that matter was continuous in structure, and this was also the opinion held by the alchemists in the Middle Ages. When modern scientific research began in the seventeenth and eighteenth centuries, the concept of atoms was revived and appeared in the writings of scientists, but it was generally mentioned incidentally and no attempt was made to use it to explain natural phenomena or to verify it experimentally.

John Dalton is regarded as the father of modern atomic theory. He was an English teacher who dabbled in chemistry as a hobby and became one of the founders of modern chemistry. The idea of atoms attracted him because he found that it could explain certain properties of gases. Later he realized that it also provided a simple explanation for the fact that elements combining to form compounds do so in fixed ratios by weight (the Law of Constant Proportions). In his book *A new system of chemical philosophy*, published in 1808, Dalton propounded his atomic theory, according to which all matter was composed of atoms, all the atoms of a particular element were identical, while the atoms of different elements differed in mass and in other properties. When two or several elements combined to form a compound, their atoms joined to form 'compound atoms', or molecules as they are called today. Dalton's ideas fell on fertile ground. At that time the experimental methods in science were sufficiently advanced to take up the challenge of testing theories on the structure of matter, and the atomic theory which was perfected by other chemists of the nineteenth century was based solidly on the science of chemistry.

The scientists of the nineteenth century succeeded in determining the mass ratios of the atoms of different elements, and to a close approximation also their absolute masses and sizes. By experiments which were both simple and ingenious, like measuring the area resulting from the spread of a drop of oil floating on water, or measuring the rate at which two different gases mix, they could establish that the diameter of an atom (and a molecule) is not much greater than, nor much less than, a hundredth of a millionth of a centimetre, or 10^{-8} centimetres in scientific notation.* This unit of length is called an angstrom (denoted Å) in memory of the nineteenth-century Swedish physicist A. J. Angström, who investigated the spectrum of the sun and the northern lights.

Faraday's researches on electrolysis at the beginning of the last century implied that the interatomic forces are essentially electric in nature, and yet, until the 1890s nothing at all was known about the internal structure of the atoms. They were regarded as tiny grains of matter, solid but elastic like billiard balls, and the name 'atom' – indivisible – was still appropriate. The first to discover anything about the internal structure of the atom was the British physicist J. J. Thomson, who is regarded as the discoverer of the electron.

The discovery of the electron

During the years 1894 to 1897 J. J. Thomson investigated the phenomenon of 'cathode rays' which had been discovered in 1858. To form these invisible rays an electric voltage was applied between two metal plates (electrodes) in a glass tube under high vacuum. The rays were emitted from the negative electrode – the cathode – and caused a glow when they impinged on the glass or a plate coated with zinc sulphide fixed inside the tube (see Fig. 1.1). Another British physicist, Sir William Crookes, had in the year 1879 postulated that the cathode rays were a stream of particles carrying negative electric

* We shall frequently use powers of ten for writing very small (or very large) numbers. 10^2 is 10×10 and 10^3 is $10 \times 10 \times 10$, etc. Negative powers are defined by the formula: $10^{-k} = 1/10^k$.

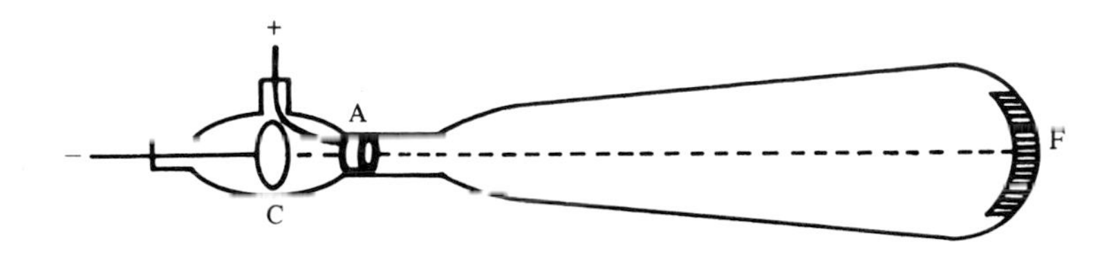

Figure 1.1. Cathode rays. A beam of electrons emitted by the cathode (C) passes through a slit in the anode (A) and produces a spot of light on the fluorescent screen (F). One can deflect the beam by applying an electric or magnetic field.

charges. Thomson verified this by showing that the rays could be deflected from their straight path by a magnetic or an electric field, and that their behaviour under the influence of these fields was exactly what would be expected of a stream of negatively charged particles. By measuring the deflection of the rays in combined electric and magnetic fields of different strengths he was able to calculate the speed of the particles as well as the ratio between the charge carried by each particle (e) and its mass (m), but no way could be found of calculating the charge and the mass separately. The ratio e/m of the particle was found to be independent of the type of metal of which the cathode was made or the residual gas in the tube. Moreover, in the year 1888 several scientists had observed that when various metals were illuminated with ultraviolet light they emitted negatively charged particles (see p. 36). Thomson repeated these experiments and found that the particles which the light knocked out of the metal surface were identical (in regard to their e/m ratio) to the particles constituting the cathode rays. Thus he reached the conclusion that these particles were present in all matter, and that by means of an electric voltage or irradiation with light they could be extracted from certain substances. The particles were given the name electrons. (Thomson was awarded the Nobel Prize in physics in 1906 for this discovery.)

When enough experimental results have been accumulated in a new area of physics, the scientists usually try to utilize these results to construct a description of the physical world, which they call a model. On the strength of Thomson's findings, the first experiment-based model for the structure of the atom was put forward in 1902. This model was proposed by the British physicist Lord Kelvin, but as it was based on Thomson's experiments and was enthusiastically supported by him, it came to be known as Thomson's model. (Since this model was later rejected in favour of a better one, its authors apparently did not insist on getting the credit for it.) According to

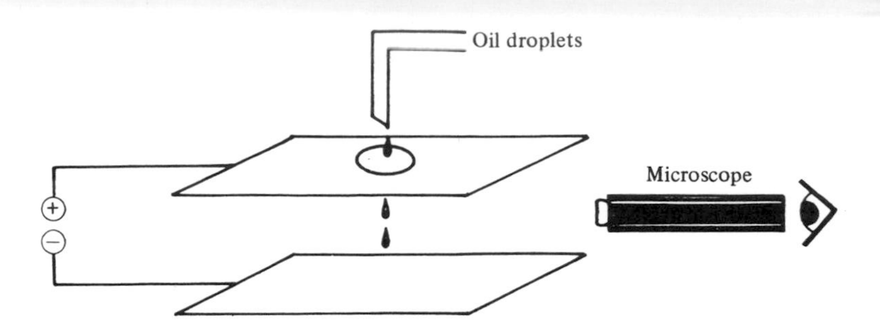

Figure 1.2. A schematic description of Millikan's experiment to measure the charge of an electron.

the model, the atom was a positively charged sphere about 1 Å in diameter in which electrons were embedded like plums in a pudding. The total charge of the negative electrons was equal to the positive charge of the sphere, and so the atom as a whole was electrically neutral. It was already known that atoms could be transformed into electrically charged bodies (ions) by various means (for example by irradiation with X-rays). According to Thomson's model, ionization of an atom (converting it into an ion) is simply the removal of an electron from it. The atom is then left with a positive charge equal in magnitude to the charge of the electron. This explanation of the ionization process provided the basis for the first estimate of the electron charge and hence also its mass (since the ratio e/m was known). The mass was found to be 2000 times smaller than that of the hydrogen atom – the lightest of the elements.

The first accurate measurement of the electron charge was made by the American physicist R. A. Millikan (Nobel laureate for 1923) in a brilliant experiment in the year 1909. By means of X-ray irradiation he charged minute droplets of oil and observed them with a microscope as they floated between two parallel charged plates (see Fig. 1.2). By measuring the speed at which the droplets fell under different electric field strengths he could calculate their charges. All the charges he measured turned out to be integral multiples of a fundamental charge whose magnitude was 1.6×10^{-19} coulombs.* It could be assumed that this charge was the charge of a single electron. From now on we shall denote the magnitude of this charge by the letter e. (The charge of the electron is $-e$, and the electron itself will be denoted e^-.) In more accurate experiments e

* The unit of electric charge is called after the French physicist Charles de Coulomb. When a wire is carrying an electric current of one ampere, a charge of one coulomb passes its cross-section each second.

was found to equal 1.6021×10^{-19} coulombs. It also turned out that the charge of most observed positively and negatively charged particles is $\pm e$. (The charge of some short lived particles is an integral multiple of e.) All measured electric charges are in fact multiples of e, but because of the minuteness of this fundamental charge scientists did not detect this fact as long as they were dealing with 'large' charges. (Even one-millionth of a coulomb contains more than a million million fundamental charges.) Millikan's success in measuring the charge of the electron made it possible to find its mass as well (since the ratio e/m was known). The mass was found to be 9.11×10^{-28} grams, or 1/1840 of the mass of a hydrogen atom.

Rutherford's experiment

In the first decade of the twentieth century several other models of the atom were proposed. These models were based on a variety of combinations of positive and negative charges, and what all of them had in common was a lack of any experimental basis. The first model based on experiment was that of the illustrious New Zealand-born physicist, Ernest Rutherford. In 1910 Rutherford's laboratory at Manchester University was engaged in research on the scattering of alpha particles in their passage through matter. Alpha particles are heavy particles – about four times as heavy as the hydrogen atom – with a charge of $+2e$, and are emitted from certain radioactive substances (see below). When a beam of alpha rays strikes a very thin metal foil, it penetrates and passes through it. In the process some of the particles are deflected from their straight path, much as bullets fired into an avenue of trees might be deflected by ricocheting off the tree trunks. Rutherford's students measured the percentage of particles deflected at different angles as the beam passed through a thin gold leaf. They placed a plate coated with zinc sulphide behind the leaf (see Fig. 1.3) and with the aid of a microscope counted the number of scintillations in a given period. Every scintillation was evidence of an alpha particle striking the plate. They then shifted the plate to a different angle and counted the scintillations in the same period. Through such measurements they hoped to learn something about the internal structure of the atom.

In these experiments only small angles of scattering were examined, because it was assumed that the heavy and fast-moving alpha particles could not change direction to any great extent on their way through the thin leaf. But one day Hans Geiger, a German researcher in Rutherford's group, was looking for a research topic for his student E. Marsden, and Rutherford suggested he might check if any of the alpha particles were deflected through large

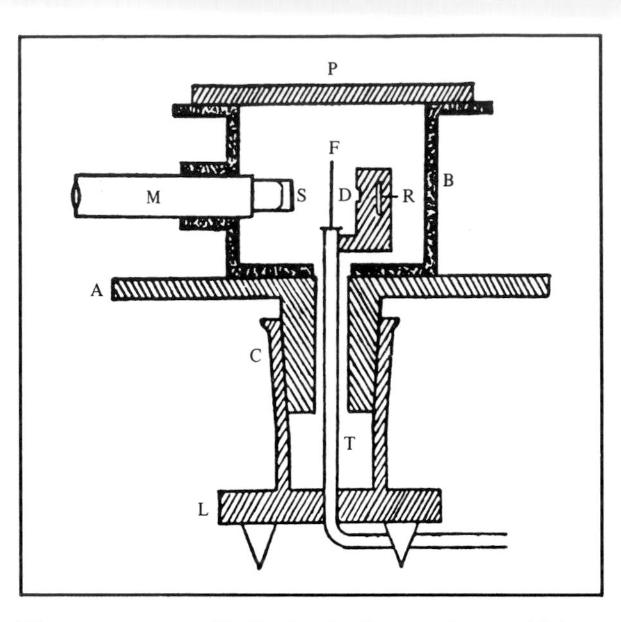

Figure 1.3. The apparatus used in Rutherford's experiment. Alpha particles from the source R, pass through the metal foil (F), and cause scintillations on the fluorescent screen (S). These scintillations are observed by the microscope (M) which can be rotated around TF. (From *Philosophical Magazine*, **25**, 604 (1913).)

angles. Rutherford did not expect any positive results from this experiment, but two days later Geiger reported in great excitement that some of the alpha particles actually recoiled back. According to Rutherford this was the most astonishing event of his entire life. In his words: 'It was almost as incredible as if you fired a 15-inch shell at a piece of tissue paper and it came back and hit you.'

Why was Rutherford so astonished? According to Thomson's model, the density of matter should be practically uniform throughout its volume, because matter was composed of closely packed spherical atoms which were themselves uniform in density. If most of the alpha particles sliced through the gold foil like a knife through soft butter, with only small deviations from their original paths, how could a few of them (about 1 in 8000) recoil like rubber balls from a stone wall?

Calculation showed that passage of an alpha particle through a single Thomson atom would cause only a small deviation. Passage through many atoms would have a cumulative effect, but it could not be expected that all the deviations would be in the same direction and would add up in the end to a large angle of deviation,

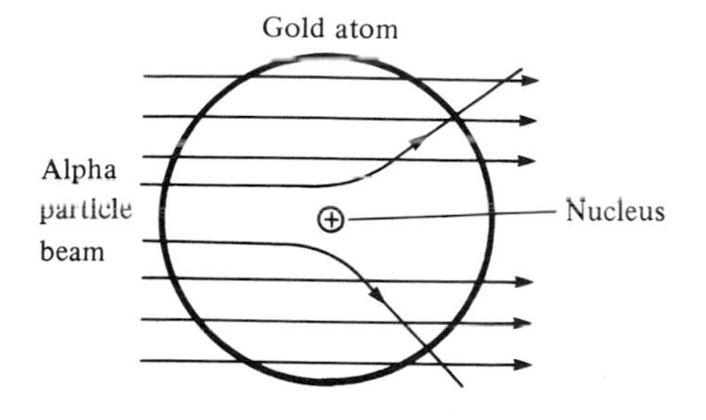

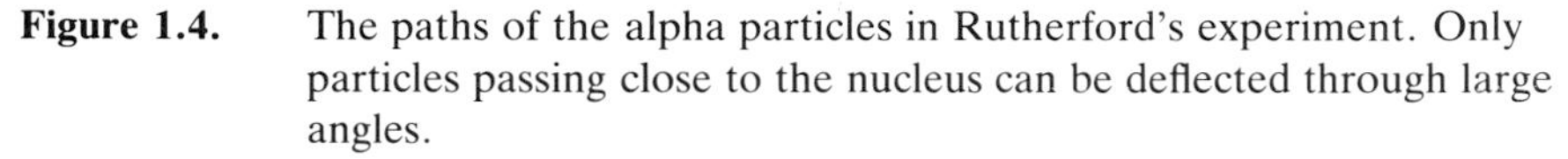

Figure 1.4. The paths of the alpha particles in Rutherford's experiment. Only particles passing close to the nucleus can be deflected through large angles.

just as it would not be expected that in a thousand tosses of a coin it would always land on the same side. There was only one explanation for the surprising phenomenon – the density of matter was not uniform! It was very dense in certain spots, and very rarefied in others.

After a series of calculations Rutherford reached the conclusion that most of the mass of the atom was concentrated in a minute nucleus which had a positive charge and a diameter only 1/100 000 of that of the atom as a whole. The negative electrons, Rutherford thought, apparently circled the nucleus like planets round the sun, and the size of their orbits was what determined the diameter of the atom. The path of the heavy alpha particles was hardly influenced at all by the light electrons but only by the heavy nucleus, which exerted an electrically repulsive force. Since the gold leaf was extremely thin, each alpha particle passed through only a relatively small number of atoms. In most cases they did not approach the minute nucleus closely and were therefore deflected only slightly, but a few of the alpha particles passed near enough to a nucleus to feel a strong repulsive force and be deflected through a wide angle (see Fig. 1.4).

Rutherford went on to develop a mathematical expression for the relative number of particles deflected through a given angle. It turned out that if his model was correct then this number should depend in a very specific way on the angle of scattering, on the charge of the scattering nucleus and on the speed of the alpha particles. The experimental set-up was improved and in a series of elegant experiments Geiger and Marsden proved that these conclusions were obeyed exactly (see Fig. 1.5).

Rutherford's experiment is a classical example of how some great discoveries are made. The recipe is as follows. When experimental

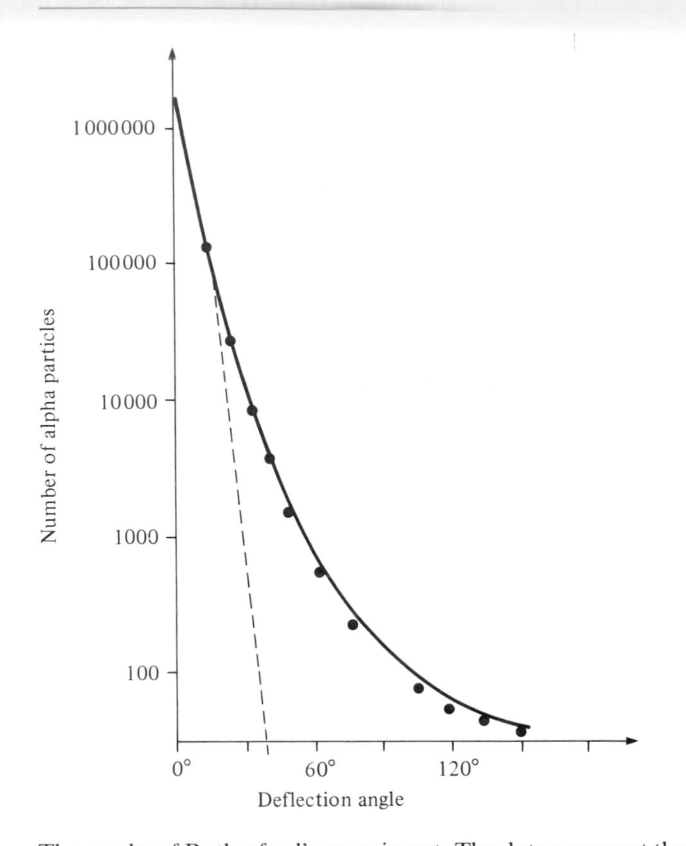

Figure 1.5. The results of Rutherford's experiment. The dots represent the measured values and the unbroken line – Rutherford's theoretical expression for the scattering of alpha particles. The agreement between experiment and theory is obviously quite good. The broken line represents the predictions according to Thomson's model of the atom. Note that most of the alpha particles experienced small deflections (only a few were deflected by an angle larger than 90°, while about a million particles were deflected by less than 10°). Nevertheless, the number of large-angle deflections was much larger than what was expected according to Thomson's model.

results do not agree with an existing model or theory, exert your brain and think up a different, more suitable one. To verify your model, draw numerical conclusions from it and test them by experiment. To this recipe one may add spice: sometimes try to test things which at first sight seem unlikely to happen.

The principles of Rutherford's model (published in 1911) of a small, heavy nucleus of radius 10^{-12} to 10^{-13} centimetres surrounded by light electrons are still valid today, although the perception of the electron orbits have changed more than once through the years, as we shall see later.

IA	IIA	IIIA	IVA	VA	VIA	VIIA	VIII			IB	IIB	IIIB	IVB	VB	VIB	VIIB	O
H 1																	He 2
Li 3	Be 4											B 5	C 6	N 7	O 8	F 9	Ne 10
Na 11	Mg 12											Al 13	Si 14	P 15	S 16	Cl 17	Ar 18
K 19	Ca 20	Sc 21	Ti 22	V 23	Cr 24	Mn 25	Fe 26	Co 27	Ni 28	Cu 29	Zn 30	Ga 31	Ge 32	As 33	Se 34	Br 35	Kr 36
Rb 37	Sr 38	Y 39	Zr 40	Nb 41	Mo 42	Tc 43	Ru 44	Rh 45	Pd 46	Ag 47	Cd 48	In 49	Sn 50	Sb 51	Te 52	I 53	Xe 54
Cs 55	Ba 56	* 57-71	Hf 72	Ta 73	W 74	Re 75	Os 76	Ir 77	Pt 78	Au 79	Hg 80	Tl 81	Pb 82	Bi 83	Po 84	At 85	Rn 86
Fr 87	Ra 88	★ 89-															

Lanthanides

*	La 57	Ce 58	Pr 59	Nd 60	Pm 61	Sm 62	Eu 63	Gd 64	Tb 65	Dy 66	Ho 67	Er 68	Tm 69	Yb 70	Lu 71

Actinides

★	Ac 89	Th 90	Pa 91	U 92	Np 93	Pu 94	Am 95	Cm 96	Bk 97	Cf 98	Es 99	Fm 100	Mv 101	No 102	Lw 103

Figure 1.6. The periodic table of elements.

Elements and isotopes

As more was learnt about the structure of the atom it turned out that the 92 elements occurring in nature could be characterized not only by the mass of their atoms – as was done by the nineteenth-century chemists – but also by the charge on their nuclei. The magnitude of the nuclear charge must be such as to balance the total charge of the electrons orbiting the nucleus. If we denote the number of electrons by Z, their total charge will be $-Ze$ (where $-e$ is the charge on a single electron). Thus the nuclear charge will be $+Ze$. Z is known as the 'atomic number' of the element and is by definition an integer.

The mathematical expression derived by Rutherford for the scattering of alpha particles made it possible to determine the value of Z of the scattering atom. At about the same time it was found that Z could also be determined from the properties of the X-rays emitted by the element* (J. J. Thomson in 1906, Moseley in 1913). It turned out that when the elements are arranged in the order of the periodic table (see Fig. 1.6), their Zs increase in consecutive numbers, i.e. 1 for hydrogen, 2 for helium, 3 for lithium, and so on. This means that the charge on the hydrogen nucleus is $+e$ and a single electron is orbiting its nucleus, the charge on the helium nucleus is $+2e$ and two electrons are orbiting round it, and so on.

This was a surprising discovery. The periodic table (or system) of the elements is a diagram in which all the chemical elements are arranged in a certain order (see Fig. 1.6). It was first proposed by the

* X-rays were discovered by the German physicist W. C. Röntgen in 1895 (in 1901 he won the Nobel prize for this discovery). These rays are generated in a vacuum tube similar in principle to the cathode ray tube (Fig. 1.1) when the fast electrons hit the anode.

Russian chemist Dmitri Mendeleev, who discovered, in about 1870, that when all the known elements were arranged according to increasing atomic weight, an interesting regularity or periodicity became apparent. This manifests itself in the remarkable similarity between elements occupying the same vertical *column* in the table. For example, all the elements of the last column are inert gases which do not tend to participate in chemical reactions (the noble gases), while the elements of the preceding column are very reactive non-metals (the halogens), and those of the first column are reactive metals (the alkali metals).

It took 40 years to go from the identification of a regularity in the properties of chemical elements to the understanding of the structural explanation. The sequence 'phenomenology–regularities–structure' is common in the evolution of science. Tycho Brahe's observations, Kepler's regularities and Newton's mechanics provide yet another example.

Note that other chemists, such as the Frenchman B. de Chancourtois in 1862 and the Englishman I. A. R. Newlands (1865) had already noticed the periodicity in the properties of elements. However, many chemical elements had not yet been discovered, and thus there were places where that ordering did not seem to prevail. Chancourtois and Newlands forced the situation and got some queer juxtapositions. Mendeleev believed in the regularity and simply left empty rubrics in his table, to await elements yet to be discovered. He also predicted the properties of the missing elements. Three of these elements were indeed discovered between 1875 and 1886 and the theory was vindicated.

The reason for the regularity was not clear until it turned out that the ordinal number of each element in the periodic table is the same as its atomic number Z! This discovery shed new light on the periodic table, implying some connection between chemical properties and the number of electrons in the atom.

The mass of the atom is conveniently expressed in units called 'atomic mass units' (amu)* (1 amu = 1.661×10^{-27} kg). The mass of the hydrogen atom is about 1.0 amu, and that of the helium atom 4.0 amu. We have already mentioned the fact that during the nineteenth century the atomic masses of the various elements were measured to some accuracy by methods based on chemical reactions. In the twentieth century more accurate measurements were performed by finding the e/m ratios of ions, using vacuum tubes similar to the one originally applied by Thomson to study the electron properties. (A modern version of this apparatus is called the mass spectrograph.) The first measurements of this sort were performed by Thomson himself. In 1913 Thomson measured the ratio e/m for ions of neon ($Z = 10$) and found that there were two kinds of neon atoms, the more abundant one having a mass of 20.0, and the rarer one having a mass of 22 (these are denoted Ne^{20} and Ne^{22}). In earlier measurements, based on chemical methods which did not distinguish between the two types of atom, the mass of the neon atom was found to be 20.18 amu. Thomson's discovery showed that this value was the 'weighted' average of the masses of the two types of atoms. (The chemists call this average the 'atomic weight' of the element.†)

* This unit is defined in such a way that the mass of one atom of the common carbon is exactly 12 amu (a small percentage of carbon atoms found in nature have a mass of 13 amu).

† It should be mentioned that *weight* and *mass* are not identical concepts. The mass measures the 'resistance' of a body to change of its velocity, while the weight is the force with which the body is attracted towards the centre of the earth. However, both can be used as a measure of the quantity of matter in the body. The mass of a body (in kilograms) is equal in its numerical value to the weight of the body (expressed in kilogram-force) when weighted near the equator at sea level.

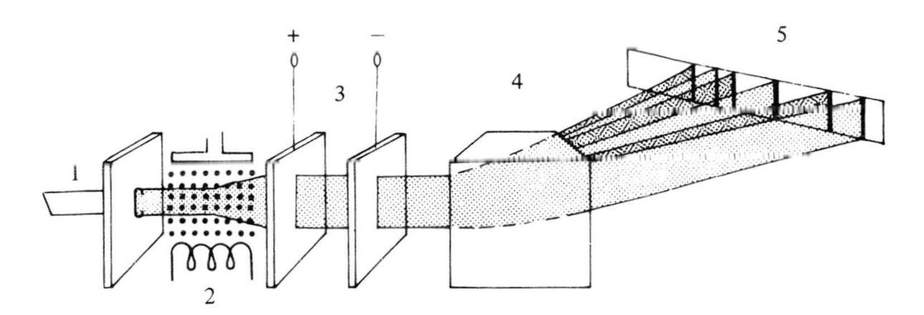

Figure 1.7. Schematic description of a modern mass spectrograph. Atoms from a source (1) are ionized by bombardments of electrons flowing from a heated cathode (2). The ions travel through an electric (3) and magnetic (4) field and eventually fall on a photographic plate (5). Ions of different masses produce different lines on the plate.

It thus turned out that atoms could be identical in their nucleus charge (Z) but different in mass. Such atoms were termed *isotopes* of the same element.

The origin of the word isotope is in the Greek *iso* meaning equal, *topos* meaning place. Different isotopes of the same element occupy the same place, or rubric, in the periodic table. Some people erroneously think that an isotope is necessarily a radioactive substance. It is true that the phenomenon was first observed, even before Thomson's experiments, in some radioactive elements, but the term isotope in itself does not imply anything radioactive. (You will read about radioactivity in the next section.) A related term is *nuclide*: all the atomic nuclei characterized by specific values of mass and charge.

In due course it was found that many elements exist as several different isotopes – all identical in chemical properties (which are determined by the number of electrons in the atom) and differing only in atomic mass. Measurements showed that the atomic mass of each isotope, when expressed in amu, is always quite close to an integer. This integer is known as the *mass number* and denoted by A. In order to understand better the definition of mass number, look at Table 1.1, which shows the atomic numbers, mass numbers and atomic masses of some isotopes of several abundant elements.

One can see, for example, that natural chlorine is a mixture of two isotopes. The lighter one (denoted Cl^{35}) constitutes about 75 % of the chlorine in nature, while the proportion of the heavier one (Cl^{37}) is about 25 %. It should be stressed that these two isotopes are chemically the same and almost identical in their physical properties, and thus their separation is not an easy task. Hydrogen has

Table 1.1. *Properties of isotopes of some abundant elements*

Element	Chemical symbol	Atomic number (Z)	Mass number (A)	Atomic mass (in amu)	Relative abundance (%)
Hydrogen	H	1	1	1.007 825	99.985
			2	2.014 10	0.015
Helium	He	2	3	3.016 03	0.000 13
			4	4.002 60	99.999 87
Carbon	C	6	12	12.000 00	98.89
			13	13.003 35	1.11
Nitrogen	N	7	14	14.003 07	99.63
			15	15.000 11	0.37
Oxygen	O	8	16	15.994 91	99.759
			17	16.999 16	0.037
			18	17.999 16	0.204
Chlorine	Cl	17	35	34.968 85	75.53
			37	36.965 90	24.47

amu = atomic mass units.

Table 1.2. *Summary of some important terms in atomic physics*

Term	Symbol	Definition
Atomic number	Z	Number of electrons in a neutral atom, or the nuclear charge (in e units)
Atomic mass		The mass of a single atom (expressed in amu)
Mass number	A	The integer closest to the atomic mass
amu		$\frac{1}{12}$ of the mass of a single C^{12} atom
Isotopes of a given element		Several types of atoms having the same value of Z, but different values of A
Nuclide		All the atoms corresponding to a single set of values of A and Z
Element		All the atoms having the same value of Z

amu = atomic mass units.

three isotopes. For most hydrogen atoms $Z = 1$ and $A = 1$; however, 1 in about 6000 hydrogen atoms is twice as heavy and has the mass number $A = 2$ (this isotope of hydrogen is called deuterium). The third isotope, tritium, with $A = 3$, does not exist in nature at all but can be produced artificially.

The various isotopes of an element are frequently designated by symbols of the form ${}_Z S^A$ or just S^A, where S is the chemical symbol of the element, A is the mass number and Z the atomic number. For example, the three isotopes of hydrogen are denoted by ${}_1H^1$, ${}_1H^2$, ${}_1H^3$. Some important terms defined in this section are summarized in Table 1.2.

In the second decade of the twentieth century the atomic number and atomic mass of many elements were measured, preparing the ground for the formulation of theories on the structure of the nucleus. All the experimental evidence pointed clearly to the fact that the nuclei of the 92 elements found in nature were composed of a small number of fundamental building blocks. Much of this experimental evidence was gathered in the course of research into the phenomenon of radioactivity, which is described briefly below.

1.2 Radioactivity

In 1896 the French physicist Henri Becquerel discovered that the element uranium ($Z = 92$) emitted a mysterious radiation which blackened photographic plates. It was named radioactive radiation. The French couple Pierre and Marie Curie discovered the same phenomenon in other elements such as thorium ($Z = 90$), polonium ($Z = 84$) and radium ($Z = 88$). The latter two elements had been unknown until then – they were discovered by the Curies in 1898. (Polonium was named after Poland, Marie Curie's fatherland.)

Becquerel, a third-generation physicist, was investigating the fluorescence of a particular uranium salt. After leaving the salt in his drawer on unexposed photographic plates, he found to his surprise that the plates were blackened. Later he proved that the uranium emitted strong radiation continuously. This fortuitous discovery is typical in science. Fleming discovered penicillin when he noticed that the bacteria in an imperfectly closed Petri dish had died, apparently killed by a mould. Others might have deduced that more care should be taken in the handling of Petri dishes. Fleming realized that this might be a very effective way of killing germs.

In 1903, the Curies shared with Becquerel the Nobel prize in physics. In 1911 Marie Curie won the Nobel prize in chemistry for discovering the elements polonium and radium (her husband, Pierre, had been killed three years earlier in a road accident). The story of the Curies, whose collaboration contributed so much to nuclear physics, casts some doubt on the wisdom of the regulations, adopted by some institutions, whereby husband and wife may not be employed in the same department . . . (There have been several such successful husband-wife teams in particle physics.)

The work of Becquerel, the Curies, Rutherford and other scientists at the end of the nineteenth and beginning of the twentieth centuries revealed the existence of three types of radioactive radiation. These were called alpha (α), beta (β) and gamma (γ) rays.

Alpha rays are particles with a positive charge of $+2e$ and a mass of about 4 amu. In 1909 Rutherford and Royds proved that α particles were the nuclei of helium atoms. (This discovery earned Rutherford the 1908 Nobel prize in chemistry.)

Beta rays are also particles, but with a negative charge of $-e$. Experiments proved that β particles were none other than electrons.

Gamma rays carry no electrical charge and have no mass. They were found to be electromagnetic radiation of very short wavelength, similar to X-rays.

In 1902 Rutherford and his colleague F. Soddy (who coined the term 'isotope', and who was the 1921 Nobel laureate in chemistry) discovered that every emission of radioactive radiation alters the emitting atom and turns it into an atom of another element – a phenomenon which they called transmutation.

For example, when a uranium atom emits an alpha particle, the emitting atom becomes an atom of thorium. In the light of the nuclear model of the atom it was therefore clear that the origin of all the radioactive radiations (including β rays) was in the nucleus, since only a change in the nucleus could transform one element into another. This conclusion was later supported by direct experimental evidence.

While emitting an alpha particle, the nucleus loses electric charge of $+2e$ and a mass of about 4 amu. The process, which is also called alpha decay, or disintegration, lowers the atomic number Z by 2, and the mass number A by 4. The equation representing the alpha decay of uranium 238, for example, is:

$$_{92}U^{238} \rightarrow {}_{90}Th^{234} + {}_{2}He^{4}$$

Remember that the number written in the upper right corner indicates the mass number A and in the lower left corner the atomic number Z. The equation tells us that uranium 238 emits an alpha particle ($_{2}He^{4}$) and transmutes into thorium 234. Note that A and Z are conserved throughout the process: the sum of the mass numbers on the right side of the equation is the same as on the left side, and this is true for the atomic numbers as well. The conservation of Z actually reflects the fact that the total electric charge is conserved. The emission of a beta particle *raises* the atomic number Z by one and leaves the mass number A unchanged (since the mass of the electron is extremely small – only 0.0005 amu). In a beta decay of lead 210 (Pb^{210}), for example, bismuth 210 ($Z = 83$) is produced:

$$_{82}Pb^{210} \rightarrow {}_{83}Bi^{210} + e^{-}$$

In this process too the total electric charge is conserved: it is 82 units before the decay and after it.

Gamma radiation is emitted from certain nuclides after emission of alpha or beta particles, and carries off excess energy beyond that which can be stably retained by the new nucleus. Since gamma radiation has no mass or charge this does not alter the charge of the emitting nucleus, the mass of which is also almost unaltered.

Other types of radioactivity were discovered in due course – it was found that some nuclides emit neutrons or positrons (see below) or split spontaneously into a number of lighter nuclei (the latter process, fission, occurs in the explosion of an atomic bomb and in nuclear reactors). An additional radioactive process, called electron capture, is that in which the nucleus captures one of the electrons orbiting around it, whereby Z decreases by one.

A turning point in research into the nucleus was the discovery that by bombarding certain elements with various types of radiation, nuclides not found in nature could sometimes be formed, among them totally new elements with Z greater than 92. The new isotopes are usually unstable and emit radioactive radiation immediately or after the lapse of some time. The natural radioactive isotopes are usually among the heavy elements at the end of the periodic table.

Artificial radioactivity on the other hand appears also in light elements. Even hydrogen, the lightest of the elements, has an artificial isotope which is radioactive – this is the tritium ($_1H^3$) mentioned above. Artificial radioactive isotopes are nowadays produced on a large scale in nuclear reactors, for use in medicine and industry. We may note in passing that the term 'artificial radioactivity' was coined in 1934 by Irene Curie, the daughter of the discoverers of radium, and her husband Frederic Joliot. They were working as a team, in the tradition of Irene's parents, and discovered that certain elements irradiated with alpha rays continued to emit radiation even after the irradiating source was removed. In earlier experiments conducted since 1919 the radioactivity ceased instantly when the inducing irradiation was stopped.

The phenomenon of radioactivity played an important role in nuclear research. Firstly, it served as a tool for bombarding the atomic nucleus – Rutherford's scattering experiments and many other studies were performed with alpha particles from radioactive sources – and secondly, the emitted radiation itself revealed much about what was going on in the nucleus.

1.3 The building blocks of the nucleus

The proton

The first experiments in artificial transmutation of nuclei were performed by Rutherford, whose name seems to be linked with every important advance in nuclear physics at the beginning of the century. In 1919 Rutherford discovered that when a sufficiently fast moving alpha particle strikes a nitrogen nucleus (N) it is sometimes absorbed and causes the instantaneous emission of a hydrogen nucleus:

$$_7N^{14} + {}_2He^4 \rightarrow {}_8O^{17} + {}_1H^1$$

The nucleus that remains behind is an isotope of oxygen (O^{17}). In this experiment Rutherford had realized the ancient dream of the alchemists – to transmute one element into another. Between the years 1920 and 1925 Rutherford and other researchers performed similar experiments with the nuclei of boron, fluorine, neon, sodium and other elements. In all cases the impact of an alpha particle caused the nucleus to emit a hydrogen nucleus. The physicists concluded that the hydrogen nucleus was one of the fundamental building blocks of all other nuclei. In view of its importance, it was accorded a special name: the proton (from the Greek word '*protos*', meaning first).

It was clear, however, that the nuclei of all the elements other than hydrogen, were not composed of protons only, since that would mean that Z would always be equal to A. Since some nuclei emit electrons (beta rays) it was natural to assume that the electron was the other building block of the nucleus. Thus the nucleus was thought to consist of A protons and $A - Z$ electrons (in addition to the Z electrons orbiting around the nucleus). This model could explain why the net charge on the nucleus was $+Z$ (in units of e) and its mass approximately A amu (the mass of the electron is 1840 times smaller than that of the proton and thus it does not contribute significantly to the nuclear mass). The helium nucleus, for example, would consist – according to this model – of four protons and two electrons. But despite the enthusiasm with which the model was received it soon ran into difficulties. Calculations showed that for several reasons such a combination of protons and electrons could not form the stable structure we know as the atomic nucleus. (At a later stage we shall discuss the physical quantity known as spin. Measurement of the spin of several nuclei proved that they could not be composed of protons and electrons.) The question thus remained: what did the nucleus contain besides protons? It was

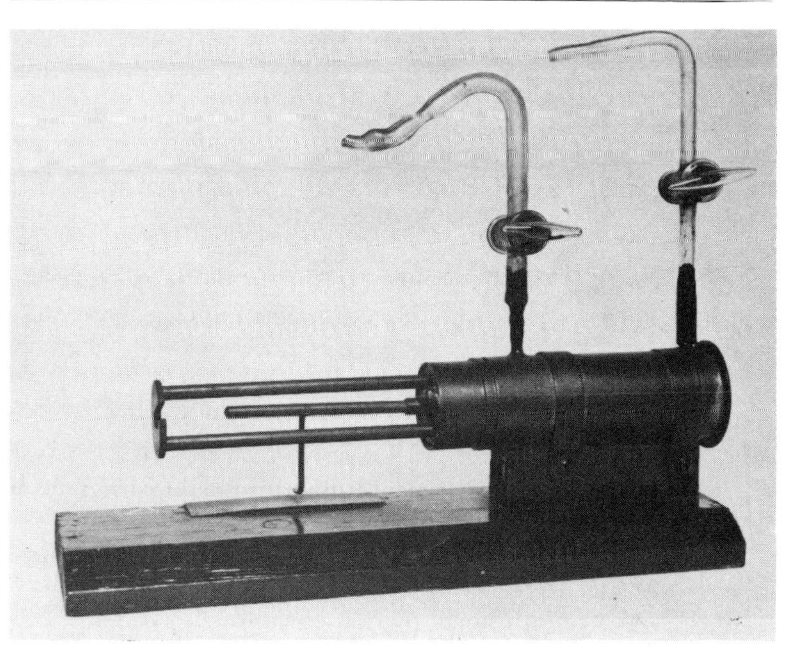

Figure 1.8. The apparatus with which Rutherford discovered the proton. (Courtesy Cavendish Laboratory, Cambridge.)

suggested that there might be an electrically neutral particle, hitherto unknown, which gave the nucleus the rest of its mass. This hypothesis was confirmed in the 1930s when the neutron was discovered.

The neutron

In 1930 two German physicists, W. Bothe (who shared, with Max Born, the physics Nobel prize in 1954) and his student H. Becker, found that when light elements such as beryllium were exposed to alpha rays, a strong uncharged radiation was emitted. They took this to be gamma radiation. But in 1932 Irene Curie and Frederic Joliot showed that if this radiation struck materials containing hydrogen, for example paraffin, it caused emission of protons – which gamma rays were never found to do. It became clear that this was a new type of particle radiation, whose properties were difficult to study because of its electrical neutrality. (Remember that the mass of the electron, for instance, was calculated from measurements of the ratio e/m.) Nevertheless, that same year James Chadwick, Rutherford's talented colleague (then at Cambridge), succeeded in determining the mass of the new neutral particle. Using the advanced equipment of the Cavendish Laboratory in Cam-

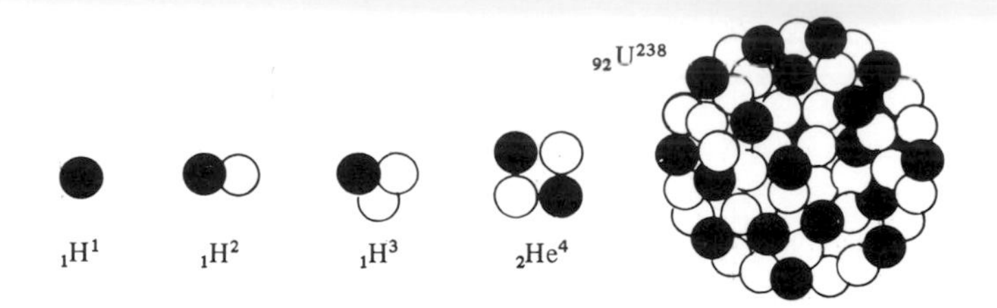

Figure 1.9. A nucleus with an atomic number Z and mass number A consists of Z protons and $A - Z$ neutrons.

bridge, which in the year 1932 was unequalled anywhere in the world, he studied the effect of the new particle on several nuclei. From the energy imparted to these nuclei Chadwick deduced that the mass of the new particle was close to that of the proton. The new particle was given the name 'neutron', and Chadwick was awarded the Nobel prize (in 1935) for discovering it.

It was soon shown that many nuclei could be made to emit neutrons, and that neutrons were also readily adsorbed by all nuclei, often causing them to split apart. It therefore seemed that neutrons were part of every nucleus, and this hypothesis was verified beyond doubt in the years that followed.

Today we know that a nucleus of mass A and charge Z is a combination of Z protons and $A - Z$ neutrons packed together. The lightest isotope of hydrogen is the only isotope whose atom does not contain any neutrons, its nucleus being one single proton. In the deuterium ($_1H^2$) nucleus there is one proton and one neutron while the nucleus of $_2He^4$ consists of two neutrons and two protons (this is also the composition of an alpha particle) and $_{92}U^{238}$ has 92 protons and 146 neutrons (see Fig. 1.9).

The discovery of the neutron is a classical example of the way in which the addition of a new building block clarifies as if by magic many previously inexplicable facts. For example, it became clear that the mass number A is just the total number of protons and neutrons in the nucleus. The fact that the atomic mass is always quite close to an integral number of amu found a simple explanation: the masses of both the proton and the neutron are close to 1 amu (see Table 1.3). Different isotopes of an element are atoms the nuclei of which have the same number of protons but not the same number of neutrons.

The neutron is represented by the letter n, and in equations it can be written as $_0n^1$ – denoting a mass of about 1 amu and a charge of zero. The proton is represented by the letter p, and in nuclear equations it is written $_1H^1$.

Table 1.3. *Properties of the atom constituents*

Particle	Electric charge (in *e* units)	Mass in kg	Mass in amu
Electron	−1	9.109×10^{-31}	0.000 549
Proton	+1	1.673×10^{-27}	1.007 28
Neutron	0	1.675×10^{-27}	1.008 67

amu = atomic mass units.

Nuclear forces

After the discovery of the neutron the question arose as to how the particles in the nucleus manage to hold together, with the nucleus consisting of such a combination of positive and neutral particles. Knowing that electrical charges of the same sign repel each other, how can the nucleus be stable?

It thus became clear that there must be some other type of force, which we shall for the time being call the nuclear force, holding the constituents of the nucleus together, that is stronger than the electrostatic repulsion between protons. Experimental support for this came in 1934 from Cambridge, where J. Chadwick and M. Goldhaber were examining the effect of gamma rays on deuterium. They found that the irradiation caused the deuterium nucleus to emit a neutron and turn into an ordinary hydrogen nucleus:

$${}_1H^2 + \gamma \rightarrow {}_1H^1 + {}_0n^1$$

This experiment gave a very accurate measurement of the mass of the neutron, and an estimate of the force that binds together the neutron and proton in the deuterium nucleus. It was later found that this binding force is the same whether it acts between neutron and neutron, proton and neutron, or proton and proton (with an electrical repulsion present as well in the latter case).

Protons by themselves couldn't constitute a stable nucleus because of the repulsive forces they exert on each other. The neutrons, because they lack electric charge, stabilize the nucleus by separating the protons from each other and by contributing attractive (nuclear) forces without exerting repulsive ones. This explains why in light nuclides such as ${}_2He^4$, ${}_8O^{16}$ and ${}_7N^{14}$ the number of neutrons approximately equals the number of protons, while in heavy nuclides such as ${}_{92}U^{235}$, ${}_{92}U^{238}$ and ${}_{90}Th^{232}$ the number of neutrons is relatively larger; as the number of protons increases each proton feels a greater repulsion due to all the other protons and more neutrons are needed to achieve stability. The phenomenon of radioactivity can be interpreted as an instability of the nucleus due to a disturbed neutron–proton balance.

Since protons and neutrons are very similar, both in mass and in their behaviour under the influence of the nuclear force, they are referred to by a common name – nucleons. They can even be regarded as a single particle which can appear in two forms – with charge or without. A neutron can change into a proton under certain conditions, and vice versa. A free neutron (outside the nucleus) turns spontaneously into a proton after an average lifetime of about 16 minutes. The transformation is accomplished by emitting an electron and another particle which will be described later. Inside the nucleus the neutron is stable and has an unlimited lifetime, except for those radioactive nuclei which emit beta rays, since every such emission involves the transformation of a neutron into a proton inside the nucleus. A free proton on the other hand, is a stable particle, but under certain circumstances a proton in the nucleus can transform into a neutron, as we shall see later.

The discovery of the neutron had a far-reaching effect on experimental nuclear physics, because the neutron was found to be a unique projectile for smashing nuclei apart. Owing to its electrical neutrality it easily penetrates nuclei even at low speeds, without being slowed down by electrical repulsion from either the nucleus or the electrons around it. When artificial radioactivity was discovered by the Curie–Joliot team in 1934 it was realized that bombardment of nuclei by neutrons was an exceptionally effective means of producing this phenomenon. The young Italian physicist Enrico Fermi found that it was particularly the slow neutrons that penetrated nuclei easily. He succeeded in making artificial radioactive isotopes of most of the known nuclei by exposing them to neutrons which had been slowed down by passage through water or paraffin. (The first experiments of this kind were performed in the goldfish pond of the research institute in Rome.) In the second half of the 1930s came the momentous discovery that when the element uranium was bombarded with neutrons its nucleus split into a number of light fragments, accompanied by the release of a large amount of energy. This process – nuclear fission – was destined to have fateful consequences – more so than all the thousands of nuclear processes discovered before it – as became clear a few years later when the mushroom cloud rose over Hiroshima.

Fermi, who had a Jewish wife, decided to leave Italy after the proclamation of Mussolini's race laws. In 1938 he was awarded the Nobel prize in physics, and took advantage of that opportunity to leave Rome with his family and continue from Stockholm to the USA. The announcement of the Nobel Committee stated that Fermi had produced the artificial elements 93 and 94. Later it was found that the interpretation of the experiments was wrong and the elements produced had actually been light isotopes (iodine and barium)

1.4 To see a particle

In the 1930s the physicist perceived the atom as being composed of particles of three kinds: electrons, protons and neutrons, each roduced by the fission of e uranium nucleus. hese elements were lentified by O. Hahn nd F. Strassmann in ermany in 1938. (The vo shared the 1944 obel prize for emistry.) Lise Meitner, ahn's other ollaborator, who as a ew had earlier been orced to leave Germany, eard about the iscovery in Stockholm. Vith her nephew, O. risch, she then iggested a model for uclear fission.

In 1942, Fermi built in Chicago the first xperimental nuclear ile. Later he articipated in the Ianhattan project at Los lamos, where the first tom bomb was eveloped. Element 100 named after him, 'ermium', and so is the nit of length used for uclear dimensions, 1 ermi = 10^{-13} entimetres.

characterized by its own mass and electric charge, as shown in Table 1.3. They managed to extricate each of them from the atom and study its properties as a free particle.

The development of suitable instrumentation plays a cardinal role in the study of sub-atomic particles. Progress often depends on the ingenuity of an experimental physicist in finding a new way of measuring some property of a minute particle. Not even the most powerful microscope will enable us to see sub-atomic particles. But there are certain instruments which so unmistakably reveal the presence of these particles and trace their paths that we can indeed claim to be able to 'see' them. Earlier we described a few experimental systems, such as Thomson's tube, with which the ratio e/m was measured. We shall now describe a number of other instruments which have served since the early years as the eyes of physicists investigating the sub-atomic world.

Most of these instruments exploit the fact that in passing through matter – whether gaseous, liquid or solid – charged particles and gamma rays strip electrons from atoms and turn them into charged ions. These ions can be detected in a number of ways. In certain substances they increase the electrical conductivity. In others they may recombine with electrons and in so doing emit light. In a gas saturated with water vapour the ions serve as centres for condensation and so the path of the particle is marked by a train of minute droplets. All these phenomena have actually been used in particle detectors. Non-ionizing particles, like neutrons, are studied indirectly. For example, neutrons are allowed to strike a substance from which they eject protons and these are then investigated.

The main instrument for studying radioactivity and X-rays at the beginning of the century was the electroscope. This consists essentially of two thin gold leaves suspended side by side inside a glass and metal container (see Fig. 1.10). When the leaves are electrically charged a repulsive force is induced between them and causes them to separate. Now when radioactive radiation or X-rays pass through the electroscope, the air molecules inside the container are ionized and the air becomes a conductor. The charge on the leaves thus passes to the earthed container walls and drains away, causing the leaves to collapse. The rate at which the leaves collapse is an indication of the intensity of the radiation.

A more convenient and accurate instrument is the *scintillator* or *scintillation* counter. One of its forms is a screen coated with a fluorescent substance which emits light when an ionizing particle strikes it. (The inner surface of a television screen is coated with a similar substance, and so are the walls of a fluorescent lamp.) In its simplest form – so successfully used in Rutherford's experiment – a

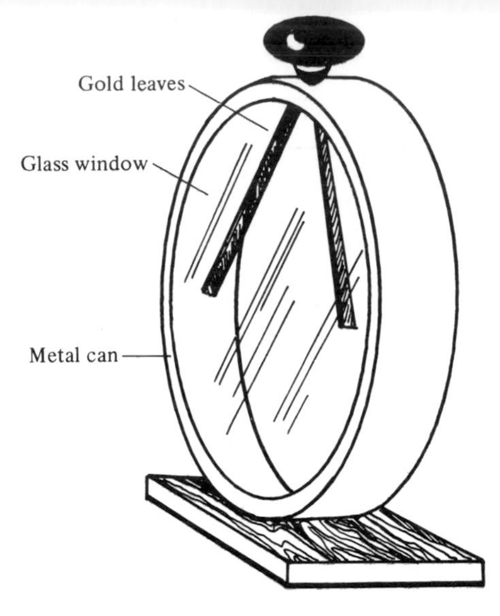

Figure 1.10. The electroscope. Two thin gold leaves are suspended inside a glass and metal box. When electrically charged particles pass through the upper metal knob, the leaves move away from each other.

microscope is placed behind the scintillation screen and the observer counts the number of scintillations in a given time. In the more sophisticated version used today the task of detecting the light is performed by an electronic tube called a photomultiplier, which converts the scintillations into pulses of electricity; these are amplified and counted by electronic circuits. Modern scintillators are made of a transparent plastic block (or liquid in a transparent vessel) placed in contact with the photomultiplier (Fig. 1.11). A particle passing through the plastic or liquid produces scintillations which reach the photomultiplier, are converted into the electronic pulses and recorded.

In another type of instrument the ionizing particle produces an electric current directly. One of the simplest instruments of this kind is the ionization chamber. This is a closed, gas-filled chamber with two electrodes to which electric voltage is applied. A particle penetrating the chamber produces ions in the gas and these are drawn to the electrodes (positive ions to the negative electrode and electrons to the positive electrode). An external circuit connected to the electrodes records the current, which is proportional to the radiation flux.

A similar but more sensitive instrument, capable of recording a single particle, is the Geiger–Müller tube. It was developed by

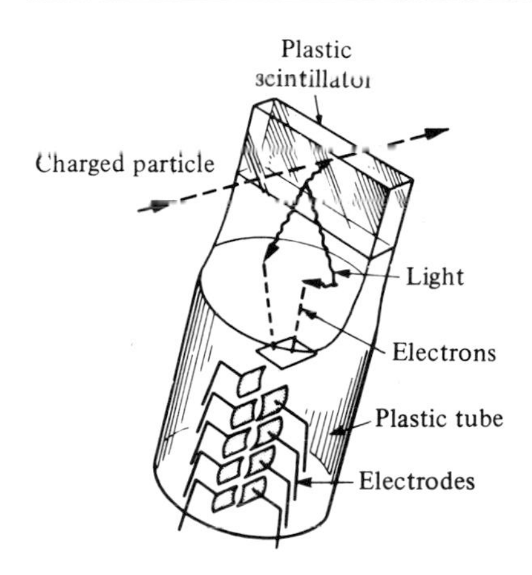

Figure 1.11. Schematic description of a scintillation counter.

Geiger in 1913 (apparently after he had tired of counting scintillations in Rutherford's laboratory . . .) and perfected by Geiger and Müller in 1928. The gas pressure and the voltage between the electrodes are such that as the ions produced by the single particle move towards the electrodes they are accelerated and ionize other atoms with which they collide, thus amplifying the current. So a single ionizing particle produces a short, strong pulse. The number of pulses reflects the number of particles that have penetrated the tube. In the 1960s, solid-state counters were developed in which the ionizing particle produces a current in a semi-conducting crystal.

Cloud chambers and bubble chambers

At the beginning of the twentieth century a simple and apparently modest invention was made which was to have a decisive influence on particle research. This was the cloud chamber, invented by the British physicist C. T. R. Wilson (Nobel prize winner for physics in 1927, together with A. H. Compton). In 1894 Wilson spent a few weeks at an astronomical observatory in Scotland, and he was impressed by the coloured halo that sometimes surrounded the shadows cast by the sun on the mountain mist. Wilson tried to produce this effect in the laboratory, and in the process discovered that tiny water droplets were formed round charged ions in the vapour-saturated air. This discovery did not arouse particular interest, until it was found that it could be used to reveal the path of an ionizing particle. In 1912 Wilson applied his invention to build a particle detector called the cloud chamber which is shown schema-

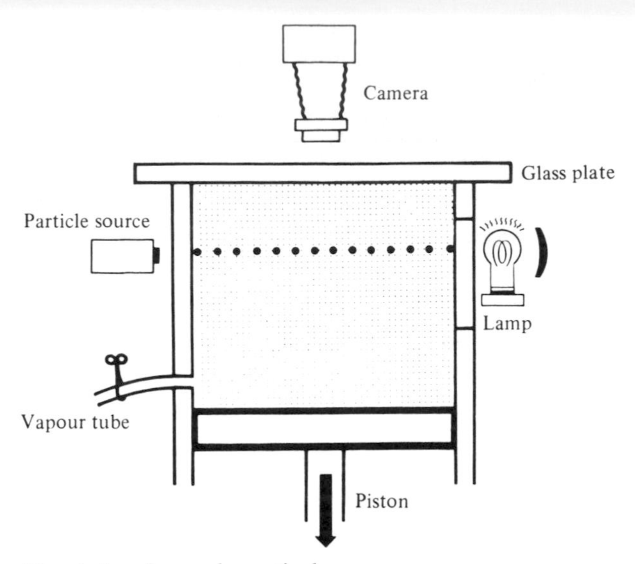

Figure 1.12. Cloud chamber, schematical.

tically in Fig. 1.12. The chamber, which is covered with glass, has a movable piston and a tube through which water or alcohol vapour is admitted. When the piston is rapidly drawn down the air expands and cools (this is the principle of the refrigerator and air conditioner), and since the amount of vapour that the air can hold decreases as the temperature drops the vapour now tends to condense into droplets. But a droplet can only form around a 'condensation nucleus' such as a dust particle. In clean air no droplets can form and a condition of 'supersaturation' develops. Now suppose that an ionizing particle passes through the chamber, leaving behind it a trail of ions. Ions are very good condensation nuclei and so droplets will condense around them, marking the path of the particle with a line of fog which can be photographed (see Fig. 1.13). From the width of the trail and its shape (a straight or zigzag line) the physicists of the beginning of the century could deduce whether the particle was an alpha or beta particle or a proton, and they even discovered new particles in this way. In later years very big cloud chambers were built – with diameters of several metres – equipped with instruments for creating magnetic and electric fields, to study the effect of these fields on the particles' motion.

In 1952 D. A. Glaser and L. Alvarez (Nobel prize winners in 1960 and 1968 respectively) developed the bubble chamber, based on a similar principle – sudden reduction of the pressure above a liquid close to its boiling point (liquid hydrogen or helium at very low temperatures, freon or propane at room temperature). The

Figure 1.13. A cloud chamber photograph, showing a beam of α particles (the two different ranges indicate that the beam consists of particles of two different energies). One of the α particles strikes a nitrogen nucleus which consequently emits a proton, whose path is seen to cross the trajectories of the α particles. (From *Proc. Roy. Soc. A*, vol. 136.)

decrease in pressure lowers the boiling point of the liquid and a situation of 'super-heating' is created, i.e. the temperature of the liquid is higher than its boiling point. If an ionizing particle passes now, bubbles of gas form around the ions, thus marking the path of the particle. (It was told that Glaser had conceived of the idea when opening a can of beer and observing the gas bubbles forming.) The first bubble chamber was only a few cubic centimetres in volume, but within a few years chambers several metres in diameter were built and they became the most widely used detectors in the search for new particles. The great advantage of the bubble chamber is: the density of the liquid is greater than that of the gas in a cloud chamber, so that the particles are more often stopped within the chamber, permitting the entire path to be photographed. Bubble chambers are easy to handle, and the particle paths can be photographed from all sides to produce three-dimensional pictures.

Scanners, usually young women, studying bubble chamber photographs obtained from large particle accelerators and translating the coordinates of the paths to numbers which would later be fed into the computer, were for years a common scene at centres of experimental particle physics throughout the world.

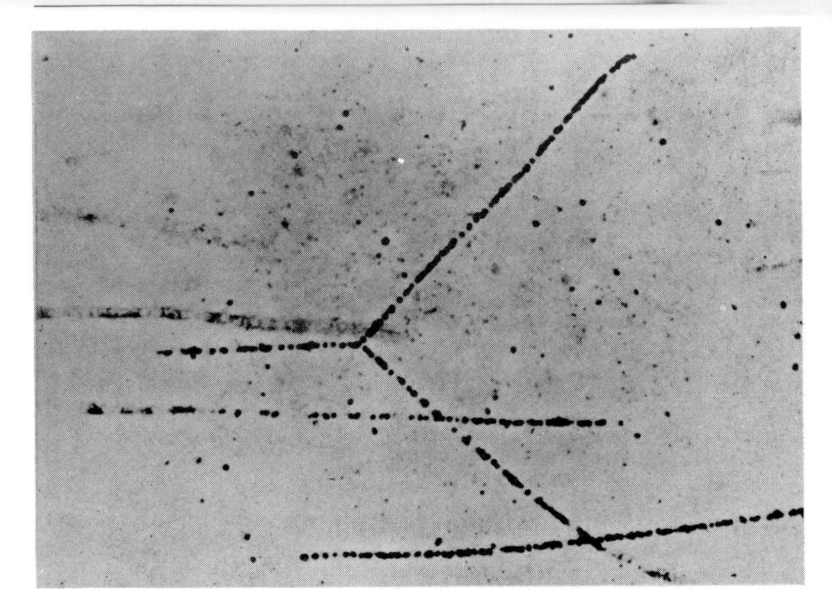

Figure 1.14. The scattering of a proton by a proton as recorded by photographic emulsion. (Courtesy C. F. Powell and Oxford University Press.)

Cerenkov counters, films and spark chambers

When a charged particle moves in a substance at a speed greater than the speed of light in that substance, light of characteristic properties is emitted. This was discovered by the Russian physicist P. A. Cerenkov (1958 Nobel prize laureate) in 1934 and became the basis for a detector similar to the scintillation counter, whose advantage is that it responds only to particles moving faster than a certain velocity.

In 1937 two Viennese ladies, Messes Blau and Wambacher, discovered that the grains of silver halide in the emulsion on the photographic film are blackened not only by light but also by fast particles striking the film. When the film is developed the path of the particle appears as a line of blackened dots (see Fig. 1.14). The path is short, since the particle is stopped much sooner in the emulsion than in air. Typical paths are only a few thousandths of a millimetre long, and are rephotographed (after the film has been developed) with the aid of a microscope. The whole process is extremely simple and does not require sophisticated equipment. Special emulsions which are much finer grained than ordinary photographic film have been developed for these studies.

A notable improvement of this research tool was achieved by the British physicist C. F. Powell and his group at the University of

Bristol. Powell was awarded the 1950 Nobel prize in physics for improving photographic emulsions and for his contributions to the investigation of cosmic rays and the discovery of the pion (see section 3.5).

Since 1959 another particle detector has come into use – the spark chamber. This chamber contains a noble gas and a number of parallel metal plates a few millimetres or centimetres apart. A sensitive counter warns of the approach of a particle beam and triggers the application of a high voltage between the plates. The particles leave trails of ions in the gas as they cross the plates, causing an electrical discharge which is seen as a spark. The path of each particle is thus marked by a series of sparks which can be photographed.

Other advanced particle detectors are described further on, when modern particle accelerators are discussed.

Chapter Two

Physical laws for small particles

Along with the exciting experiments that changed the face of physics in the twentieth century, work proceeded feverishly on the theoretical side to rebuild the conceptual world that had been severely shaken. Sometimes the theorists lagged behind the experimentalists, and certain experimental results waited for several years to be explained. And sometimes experiment lagged behind theory, and a new theoretical model waited for years to be verified experimentally. Two great theories left their mark on physics in this century: the theory of relativity, and quantum mechanics. In this chapter we shall briefly review the main ideas of these theories without which the atomic world cannot be understood.

2.1 The theory of relativity

The theory of relativity was formulated by the great physicist Albert Einstein. The first part of it (and the important one for us) – the special theory of relativity – was published in 1905, and sprang from the need to explain a puzzling question that exercised the physicists of the nineteenth century: why was it impossible to alter the speed of light, which remained constant (299 792.5 kilometres per second) even when the detector was moving relative to the source of light? In contrast to other physicists, Einstein did not try to explain this phenomenon. He rather accepted it as an established fact, and drew far-reaching conclusions which, on the face of it, seem to go counter to common sense. One of the important conclusions was that the speed of light in a vacuum is an upper limit to speed. Nothing can move faster, and no mass can even reach this speed. Another interesting conclusion was that the mass of a body increases when its speed increases, according to the formula

$$m = \frac{m_0}{(1 - v^2/c^2)^{1/2}}$$

where m is the mass of the body, m_0 is its mass when at rest (its 'rest mass'), v its speed and c the speed of light (see Fig. 2.1). The speeds which can be reached in our daily life are small compared with the speed of light and therefore we cannot detect the phenomenon of 'mass swelling' in everyday life. But in the world of small particles speeds close to the speed of light are quite common. As early as 1897 it was found that the ratio e/m of the electrons in the cathode tube decreased when their speed, which increased with the voltage across the tube, approached the speed of light. After publication of special relativity it became clear that this was due to the increase of m. A conclusion which is even harder to get used to is that the flow of time changes at high speeds. When a particle moves at a speed close to that of light, its 'internal clock' moves more slowly. This conclusion too was fully verified by experiment: short-lived particles that were accelerated to high speeds disintegrated after a longer time than identical particles moving slowly.

Mass and energy

From what has been said above it follows that when energy is imparted to a body and its speed is increased – its mass increases. But the connection between energy and mass is much more general.

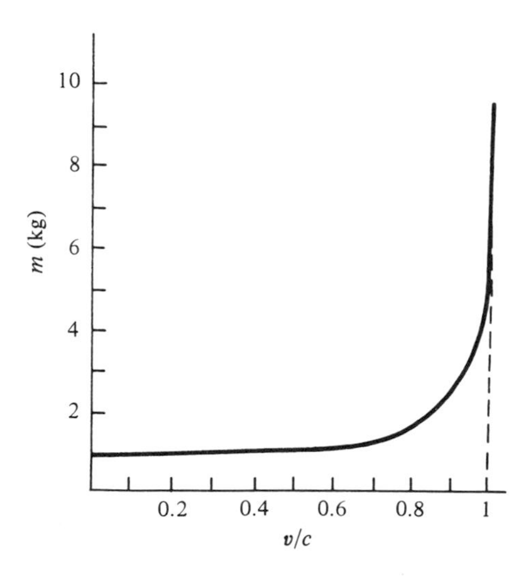

Figure 2.1. The mass (m) as a function of the ratio between the velocity (v) of the body and the speed of light (c).

Energy can be transformed into mass and mass into energy, according to the famous formula

$$E = mc^2$$

where m is the mass in kilograms, E the energy in joules and c the speed of light in metres per second. When this formula was first presented by Einstein it seemed to be a solely theoretical exercise, but today the conversion of mass into energy is a familiar process. This process occurs continuously in nuclear reactors – the mass of the 'nuclear fuel' in the reactor decreases as energy is produced from it. The same process occurs in a nuclear explosion: were it possible to collect all the fragments of a nuclear bomb after it had exploded, we would find that their total mass was less than that of the whole bomb before the explosion. The difference in mass is converted into the destructive energy of the bomb. One can get an idea of the enormous amount of energy locked up in mass from the fact that one gram of matter converted entirely to energy liberates 25 million kilowatt-hours, about the amount of electricity that 2000 thrifty families use in 6 years.

Einstein's energy formula is very important in the fields of nuclear physics and elementary particles. If we look at a process in which particles react to produce other particles, the mass of the particles produced is not necessarily equal to the mass of the particles entering the process. But if both the mass and the kinetic energy of the particles are taken into account, it is found that the sum of energy + mass (expressed in the same units) is conserved. Physicists sometimes express the mass of a particle in units of energy. A unit frequently used for this purpose is the electronvolt (eV), which is the energy gained by an electron when it is accelerated by a potential difference of 1 volt. One electronvolt is equal to 1.6×10^{-19} joules. (The energy of electrons in an X-ray tube is several times 10 000 eV). It is convenient to express the masses of the elementary particles in MeV (million electronvolts) or GeV (10^9 eV). The rest mass of the electron, for example, is 0.51 MeV and that of the neutron and proton 939.5 and 938.2 MeV respectively; 1 amu is equal to 931.5 MeV.

---

1 eV = 1.6×10^{-19} joules
1 MeV (=million eV) = 10^6 eV
1 GeV (='Giga' eV) = 10^9 eV

Accurate measurements show that the mass of a nucleus is always smaller than the sum of the masses of the nucleons of which it is composed. Table 1.3 (p. 19) shows that the mass of two protons and two neutrons is 4.032 amu, whereas the mass of the helium nucleus which is composed of these four nucleons is only 4.0015 amu. If we could form a helium nucleus by combining its four constituents we would find that in the process an amount of energy equivalent to the mass difference – 0.03 amu or about 28 MeV – was liberated. (Processes of this kind occur within the sun and other stars, and also in the explosion of a hydrogen bomb.) In order to break the helium nucleus down again into four separate nucleons we would have to supply at least that amount of energy to the nucleus. This energy is therefore called the binding energy of the nucleus.

A dramatic demonstration of Einstein's equation is the process called 'pair production': a gamma ray particle, which has zero rest mass, is transformed in the vicinity of a nucleus into two electrically charged particles of matter. On the face of it, mass has been created out of nothing, but in fact the energy of the gamma ray has been transformed into the mass of the particles. This process will be discussed in greater detail later.

2.2 The split personality of light and the birth of quantum mechanics

Quantum mechanics, like the theory of relativity, was developed to explain phenomena which could not be explained within the framework of the accepted 'classical' physics. Some of these phenomena relate to the properties of light, and others to the structure of the atom.

What is light?

In the past 250 years physicists have given various answers to the question: 'What is light?'. In 1704 Newton, in his book *Optics*, described light as a stream of particles or corpuscles. At about the same time the Dutch scientist Christiaan Huygens suggested that light was a wave, but because of the unchallenged authority of Newton this idea was rejected in favour of the corpuscular theory of light. The concept of a wave is easily understood in terms of a simple example – the waves set up in still water when a stone is dropped into it. Fig. 2.2 represents schematically the surface of the water at a particular instant. The maximum height of the wave is called its amplitude. The distance between two identical points (for example between two crests) is called the wavelength (λ). The frequency of the wave (f) is the number of times a point on the surface of the water rises and falls in one second. The product of the wavelength and the frequency gives the speed of the wave:

$$v = \lambda f$$

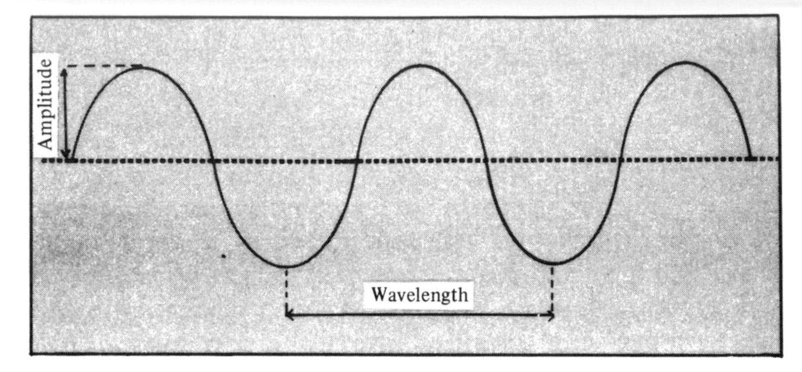

Figure 2.2. Waves. A schematic description.

The wave carries energy with it in the direction of travel, but it does not carry matter with it since the movement of the surface of the water is up and down and not forwards. (The movement of the big waves in the ocean might be a little different.) Such a wave – in which the medium transmitting the wave moves at right angles to the direction of travel, is called a transverse wave. Sound waves are of a different type, called longitudinal waves: as the wave propagates through the air, the air molecules oscillate backwards and forwards in the direction of travel of the wave.

To return to the problem of light. In 1803 the English physicist Thomas Young performed an experiment which proved conclusively that light was a wave. Young showed that when a screen was illuminated with two beams of monochromatic light there appeared on the screen – under certain circumstances – dark and bright lines (see Fig. 2.3). These lines, called interference patterns, cannot be accounted for by the corpuscular theory of light. However, they are easily explained by the wave theory. When two waves of the same wavelength reach a certain point they may cancel each other. This will happen if the waves are opposed to each other in such a way that when the one wave is at its maximum the other is at its minimum and vice versa. If, however, the maxima of the two waves coincide, the minima will coincide too and the waves reinforce each other. This phenomenon can be seen if two stones are thrown into still water. Each stone will set up spreading circular waves around itself. When the two wave systems meet there will be spots where the water does not move at all, and spots where the height of the waves will be doubled. The interference phenomenon which Young demonstrated therefore proved beyond doubt that light was a wave. The phenomenon of polarization of light (on which 'polaroid' sunglasses are based) proved that light was a transverse wave.

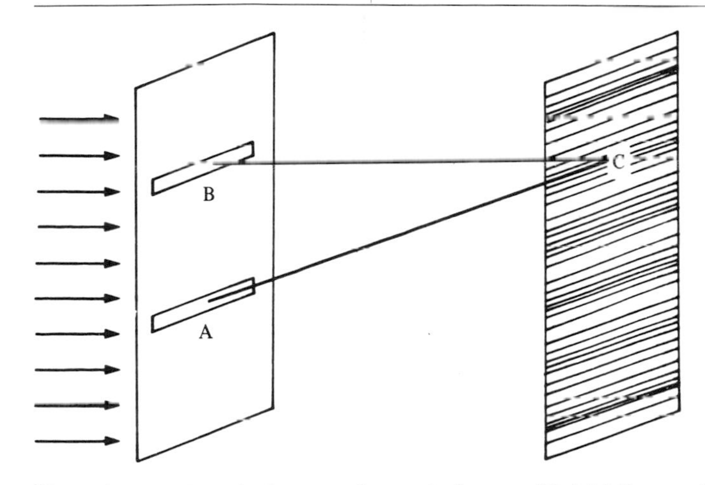

Figure 2.3. Young's experiment. A monochromatic beam of light falls on a barrier with two slits (A, B). Each of the slits behaves as a monochromatic light source, resulting in a typical interference pattern of dark and light bands on the confronting screen (C is an arbitrary point on the screen).

At the end of the last century the Scottish scientist James Clerk Maxwell formulated four equations which accounted for all the known electric and magnetic phenomena. Analysis of Maxwell's equations showed that if an electric charge oscillated it would set up waves which were in fact oscillations of the electric and magnetic *fields* surrounding the charge. These waves transmitted energy, which should be detectable. In 1888 Heinrich Hertz in Germany succeeded in producing such electromagnetic waves and detecting them with a receiver some distance away, thus providing experimental confirmation of Maxwell's theory. (The unit of frequency is named after Hertz: 1 hertz = 1 cycle per second.) The astonishing fact was that the speed of these waves in a vacuum, according to Maxwell's equations, was identical to the speed of light as measured experimentally. This indicated that light itself was actually an electromagnetic wave. The different types of electromagnetic waves – radio waves, microwaves, ultraviolet light, visible light, infrared radiation, X-rays and gamma rays – differ only in their frequency and wavelength (see Fig. 2.4). The wavelength of visible light determines its colour (violet light has a wavelength of 4×10^{-5} centimetres and red light 7×10^{-5} centimetres).

Black body radiation

It seemed as if the riddle of light had at last been solved. But a few phenomena – which seemed trivial at first sight – remained unexplained. One of these was thermal or 'black body' radiation –

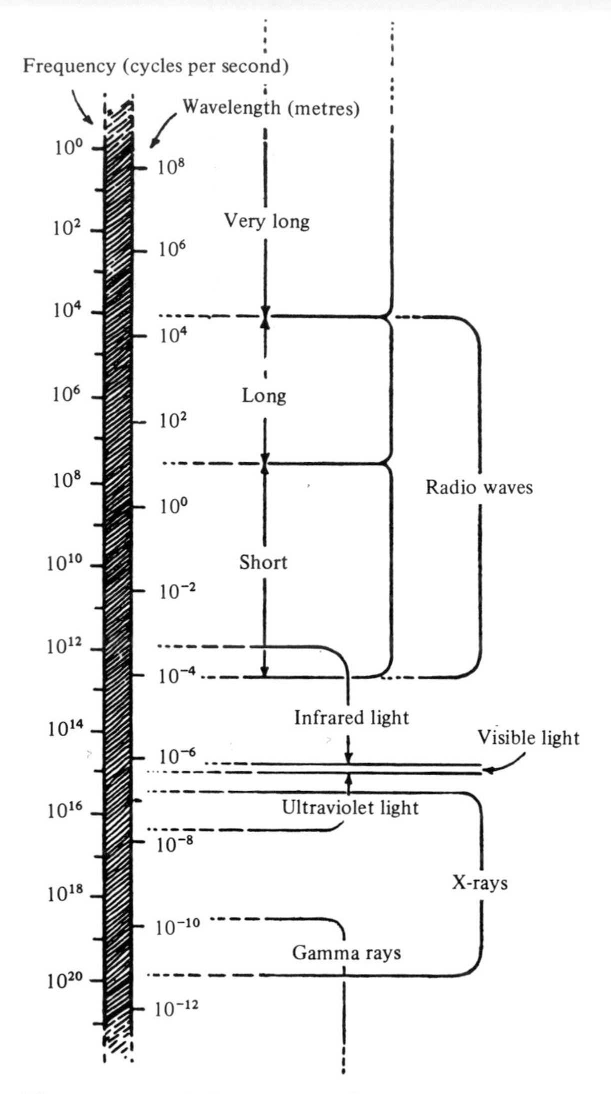

Figure 2.4. The spectrum of electromagnetic waves.

the radiation emitted by bodies when they are heated. In the last century, physicists investigated the spectrum of this radiation, that is, they measured the intensity of the radiation emitted at different wavelengths. They found that the spectrum changed with temperature, but was not affected at all by the nature of the substance emitting the radiation. The incandescent filament of a lamp at a certain temperature radiates exactly the same light whether it is made of steel, tungsten or silver.

Attempts to explain this spectrum in terms of classical physics were a total failure – according to the classical laws the spectrum should have been very different from what was actually observed. It was the German physicist Max Planck (winner of the physics Nobel prize in 1918) who solved the mystery in 1900. He found that the discrepancy could be avoided by revising the accepted assumption that light from a radiating body was emitted in a continuous way, like water flowing from a tap. Let us suppose, said Planck, that light is emitted and absorbed by matter only in definite discrete amounts, which Planck called quanta (plural of quantum). A body cannot emit or absorb light energy in any arbitrary amount but only in integral multiples of the basic amount or quantum. Planck showed that with the aid of this simple assumption it was possible to get a mathematical expression which accurately described black body radiation. By comparing theory and experiment it was found that the energy E of a quantum should be proportional to the frequency of the light f, and inversely proportional to the wavelength λ:

$$E = hf = hc/\lambda$$

The proportionality factor h is known as 'Planck's constant' and is one of the most important constants in the quantum theory and in modern physics in general. The value of h is 6.63×10^{-34} joules · second.

Despite the impressive success of Planck's model in explaining the spectrum of the black body radiation, most physicists refused to accept it at first. And no wonder. The concept of quanta ran counter to the accepted wave theory of light which fitted almost all the observations so well. It rather suggested the reinstatement of the corpuscular theory which had already been rejected. It was not clear how one could reconcile the concept of wave behaviour which was so nicely demonstrated by interference phenomena, with the particle behaviour expressed here. Altogether it was an ad-hoc hypothesis which could not yet be fitted smoothly into the known physics of light. This situation lasted some 25 years, until the advent of quantum mechanics in 1925.

Waves and particles

Albert Einstein was the first to show that Planck's assumption could explain other incomprehensible physical phenomena as well. In 1905, in an article in the German journal *Annalen der Physik*, Einstein explained the photoelectric effect with the aid of the concept of quanta. (This article won him the Nobel prize in 1921. In the same year there appeared four other articles by Einstein, who was then a clerk in the Swiss Patents Office. One of these was a summary of the special theory of relativity, and another proposed the idea that mass could be converted into energy and vice versa.)

What is the photoelectric effect? When light strikes metal, the metal emits electrons, on condition that the wavelength of the light is shorter than a certain wavelength, λ_0, which depends on the metal. (This phenomenon is the basis of the photoelectric cell, which prevents an elevator door from closing when someone is standing in the doorway, cutting off the beam of light, and thereby the cell's current too). Potassium, for example, emits electrons when it is irradiated with violet or ultraviolet light. Irradiation with light of longer wavelength, for example yellow or blue, does not cause emission of electrons, no matter how strong the intensity of light. But, the violet light causes emission of electrons from potassium even when its intensity is very weak. This was completely incomprehensible until Einstein's article appeared.

Einstein went further than Planck and assumed that quanta of energy exist in a beam of light naturally, and not only during the process of emission and absorption. A beam of light, therefore, is a stream of particles – which Einstein called photons – with no mass but having an energy of hf per photon. In order to liberate an electron from a metal, it must be given a certain amount of energy. This is a sort of 'travel tax' that must be paid in order to cut the electron loose from the metal, which binds it with a certain energy. This requires a photon whose energy exceeds the 'travel tax'. If the irradiating light has too long a wavelength (too low a frequency) it will be of no avail to increase the intensity – which means simply increasing the number of photons – since each individual photon is too weak to knock out an electron. (At ordinary light intensities it would be highly unlikely for two photons to hit the same electron at the same time.) There is a simple analogy for this: it would be impossible to breach the armour plating of a tank with a volley of bullets, however intense, but if one has an anti-tank shell, a single shot will suffice.

Thus the photoelectric effect could be explained by assuming that light was a stream of particles, but what about Young's experiment and the many other experiments which proved beyond question that light was a wave? The realization gradually permeated physics that light had the properties of both waves and particles. Under certain conditions it revealed its wave nature and under others its particle nature. Even experienced physicists found it hard to digest this idea which conflicted with all accepted notions. The question that recurred naturally was: Yes, but what is the 'true' nature of light – particle or wave? The answer seems to be that the familiar mental images at our disposal, which are drawn from our experience with 'big' or macroscopic bodies, are incapable of giving us a concrete picture of light that will be universally applicable to all circumstances. We are therefore forced for the time being to be satisfied with two models which each fit certain circumstances. It is reassuring, however, to note that the difficulty arises only in regard to the concrete visualization of the model, and not in the mathematical theory. The complete physical theory of light has built-in provisions for all circumstances in one consistent system, and determines clearly the conditions under which the particle nature or the wave nature will predominate. We shall see later that this duality is not confined to light only, but that the matter particles too suffer from this apparent 'split personality' in our concrete visualization of them.

The Compton effect

During later years further evidence of the dual nature of light was accumulated and it was found to be a virtue not only of visible light but of all the electromagnetic spectrum. An important manifestation of this duality was discovered in 1922 by the American physicist A. H. Compton (physics Nobel laureate for 1927 together with Wilson, the inventor of the cloud chamber). Compton found that when a block of paraffin was irradiated with X-rays, a part of the scattered rays emerged with wavelengths longer than the original one. The effect could not be explained by the wave picture of the electromagnetic radiation. Compton, however, found that it could be explained by the photon model, if it was assumed that the photons collided with electrons in the material and transmitted energy to them, exactly as would occur in the collision of billiard balls.

One of the most important quantities defined in classical mechanics is the momentum. The momentum of a body is the product of its mass and velocity. The importance of this quantity stems from the fact that the momentum of a system is conserved as long as the system is not acted upon by an external force. It was found that although photons do not have mass, they do carry momentum. Compton's experiment has shown that a photon of electromagnetic radiation of wavelength λ carries momentum whose value is $p = h/\lambda$, where h is Planck's constant. Compton explained that in the collision with an electron the photon loses momentum, and therefore its wavelength increases by the scattering.

The spectrum of the black body radiation, the photoelectric

effect, Compton's experiment, as well as many other phenomena have proven beyond any doubt the dual nature of the electromagnetic radiation and the relations

$$E = hf, \ p = h/\lambda$$

between the frequency (f) and wavelength (λ) of the wave, and the energy (E) and momentum (p) of a single photon.

2.3 Bohr's model of the atom

The photon model actually marked the birth of what later came to be called the 'old' quantum theory. The main motivation for the further development of the theory was the difficulty in elucidating the structure of the atom.

Rutherford assumed that the electrons orbited round the nucleus like planets round the sun. But this model had a serious drawback. A charged particle moving in a circle must, according to Maxwell's equations, emit electromagnetic waves, and in so doing it must gradually lose energy. The speed of the electron should therefore decrease, causing it to get closer and closer to the nucleus till they eventually cling together. This is in conflict with the fact that atoms are stable and do not collapse.

In 1913 the Danish physicist Niels Bohr (Nobel laureate for 1922) proposed a brilliant solution. It had already been shown by Planck and Einstein in their works on 'particles of light' that the microscopic world of small particles is governed by laws which differ from those of the macroscopic world of large bodies. Let us suppose that the electron in the atom is not free to move in any arbitrary orbit but must obey a law which permits only certain circular orbits. So long as the electron remains in its orbit it does not radiate energy. However, if its orbit is not the innermost one it can jump to an orbit closer to the nucleus, emitting a quantum of energy (see Fig. 2.5). That in brief is the principle of Bohr's atom.

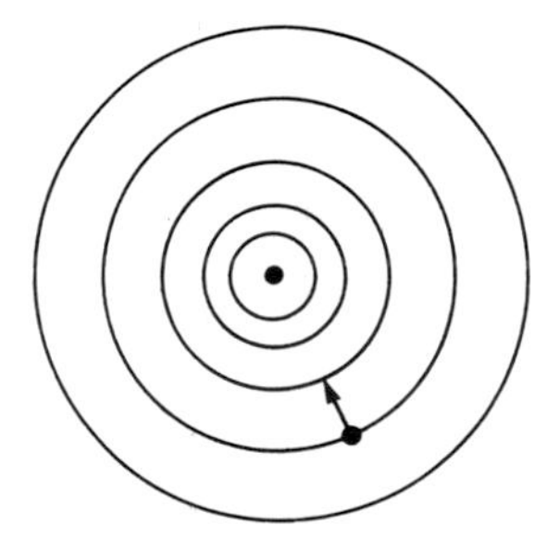

Figure 2.5. According to Bohr's model, the electron in the hydrogen atom is confined to certain circular orbits. Each transition between orbits is accompanied by the emission or absorption of a photon.

How would the rule of 'allowed orbits' be formulated? Just as energy can only be transmitted in quanta, let us suppose that the *angular momentum* (see section 7.4) of the electron in the atom must be an integral multiple of a certain natural unit. This forces the electron to move only in orbits with certain specific radii. There is a minimum angular momentum, which determines the closest orbit to the nucleus, and closer than this the electron cannot get.

This model was highly successful in explaining the properties of the hydrogen atom. It was found that if one assumes that the angular momentum of the electron is an integral multiple of $h/2\pi$ (where h is Planck's constant), the model accounted very accurately for the wavelengths of light that the hydrogen atom was observed to emit or absorb. This 'quantization rule' fitted nicely into the photon model: just as light energy can be delivered only in integral multiples of hf, the angular momentum of a system must be an integral multiple of a 'natural unit' which equals $h/2\pi$ (this unit is denoted $\hbar$).

As time went on it became clear, however, that the initial acclaim for Bohr's model was premature. Attempts to elucidate the structure of other atoms on the basis of this model failed. The model could also not explain phenomena such as the influence of a strong magnetic field on the light emitted by certain atoms. It became clear that Bohr's model was still only an ad-hoc theory and that a more comprehensive theory was needed in order to explain the atomic structure.

2.4 de Broglie: the electron is a wave

That comprehensive model for the behaviour of the electrons in an atom did not spring up all at once but was built up in stages by a cooperative effort on the part of physicists from all over the world who finally produced an entirely new theory of sub-atomic particles. One cornerstone was laid by Prince Louis de Broglie, a member of an aristocratic French family, who took up physics under the influence of his older brother, Duke Maurice, who was himself a well known physicist. In 1923, under the guidance of Langevin, who had consulted Einstein, de Broglie presented his doctoral dissertation to the French Academy of Sciences on 'The connection between waves and particles'. The young de Broglie, who at 31 had served in the army for 6 years during World War I, presented in a series of lectures a revolutionary idea which was described by another scientist as the Second French Revolution. Just as light has particle properties, said de Broglie, so material particles must have wave properties. Every moving particle has an associated wave of definite wavelength and frequency determined by the mass and velocity of the particle. The laws of motion of small particles cannot be understood unless the wave nature of the particles is taken into account, just as the photoelectric effect and black body radiation cannot be understood without resort to the particle properties of light.

The mathematics of this model was simple. de Broglie assumed that the equations $E = hf$ and $p = h/\lambda$ were valid for material particles as well as for photons. Thus, the wavelength, λ, of the particle is given by $\lambda = h/p$ where p is its momentum. The faster the particle moves, the shorter is its wavelength.

de Broglie's daring idea was not based directly on any experimental result – unlike the photon model of Planck and Einstein and the theory of relativity, for example. But in 1927, 4 years after it was published, the idea gained some very convincing experimental support. Clinton Davisson and Lester Germer, two young physicists at the Bell Telephone Co. in the USA, investigated an electron beam reflected from a polished metal surface and found in the reflected beam maxima and minima such as in X-rays reflected from a crystal. Still more convincing evidence came shortly after from George P. Thomson of England, who passed fast electrons through a thin layer of material and got bright and dark rings on a photographic plate. There is no explanation for this if the electrons are regarded as 'classical' charged particles. In that case the photographic plate should have been uniformly blackened. But the phenomenon can be clearly explained if the electron beam is assumed to have a wave character, since, as we have seen, waves can reinforce each other in certain places and cancel each other elsewhere. In fact, X-rays passing through a crystal form an identical pattern of rings. In 1929 de Broglie won the Nobel prize for physics in recognition of his model, and in 1937 the prize was shared between Davisson and Thomson who confirmed the model.

George P. Thomson is the only son of J. J. Thomson, the discoverer of the electron. He studied at Cambridge University and later carried out his research there under the guidance of his father. It is the irony of fate that the father was awarded the Nobel prize for proving that the electron was a particle, and the son received the same prize, 31 years later, for showing that the electron was a wave.

The concept of wavelike properties of material particles is even harder to digest than the particle-like nature of light. You ask yourself: 'If I could look through an imaginary microscope at an electron moving in a straight line, what would its associated wave look like? Would it express itself in a constant vibration of the particle? Or is the wave a kind of invisible field floating round the particle?' The answer is that any such visualization is erroneous. The best way of thinking of the particle is as a small body moving exactly like a 'classical' particle unless a force acts on it or it collides with another particle, in which case it moves not as a classical particle but according to the equations resulting from its wave character. The wave associated with the electron is not a tangible physical quantity (actually it is a complex number). However, it controls the electron's motion, which differs from the motion of a 'classical' body under the same circumstances.

Large and small bodies
Over the years the wave nature of matter particles was proved beyond doubt. It was shown that not only electrons but also protons and neutrons, as well as atoms and molecules, behave like waves. It was found that the wave property is actually common to all bodies, but in large bodies it cannot be detected, since the waves associated with them are of very short wavelength and there is no way of measuring or observing them. Let's demonstrate it by an example. The wavelength of an electron accelerated through a voltage of 10 000 volts is 0.12 Å (see calculation below). This is about the wavelength of 'hard' X-rays – and indeed when such an electron beam passes through a crystal an interference pattern similar to that of X-rays is obtained. But, the wavelength of a tiny dust particle with a mass of one-millionth of a gram, moving at a speed of 1 metre per second, is only 6×10^{-15} Å (and as the speed increases the wavelength becomes shorter). There is as yet no way of measuring such a short wavelength.

The kinetic energy of an electron accelerated through 10 000 volts is 10 000 electronvolts, i.e. 1.6×10^{-15} joules. Using the formula for kinetic energy we have:

$$\tfrac{1}{2}mv^2 = 1.6 \times 10^{-15}$$

Substituting the electron mass, $m = 9.1 \times 10^{-31}$ kilograms, we find $v = 6 \times 10^7$ metres per second. Hence the electron wavelength is:

$$\lambda = \frac{h}{mv} = \frac{6.6 \times 10^{-34}}{9.1 \times 10^{-31} \times 6 \times 10^7} = 0.12 \times 10^{-10} \text{ metres}$$

$$= 0.12 \text{ Å}$$

(At higher speeds the relativistic formulae for energy and momentum must be used.)

The wavelength of a tiny dust particle of mass one-millionth of a gram (10^{-9} kilograms), moving at a speed of 1 metre per second, is:

$$\lambda = \frac{h}{mv} = \frac{6.6 \times 10^{-34}}{10^{-9} \times 1} \approx 6 \times 10^{-25} \text{ metres} = 6 \times 10^{-15} \text{ Å}$$

2.5 Quantum mechanics

How did de Broglie's model contribute to a better understanding of the world of small particles? de Broglie himself showed how his

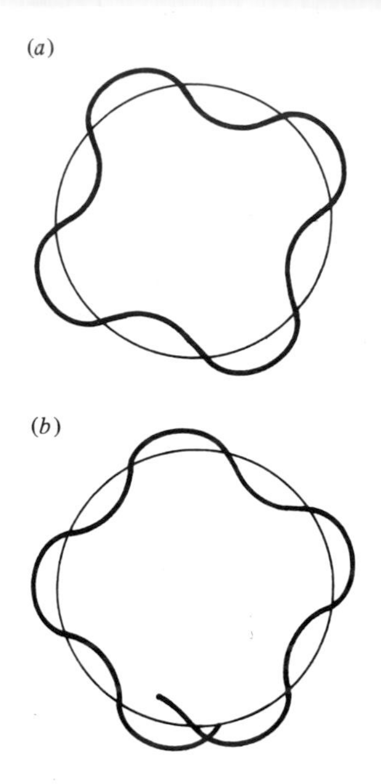

Figure 2.6. If the electron is a wave, the number of wavelengths along the circumference of its orbit must be an integer (*a*), otherwise the wave would obliterate itself by destructive interference (*b*).

model provided a conceptual foundation for Bohr's ad-hoc quantization law. He assumed that the electron indeed orbited the nucleus, at a speed v in a circle of radius r. But since the electron was also a wave, the circumference of the orbital circle had to be an integral number of wavelengths (Fig. 2.6*a*) otherwise the wave would interfere destructively with itself and obliterate itself (Fig. 2.6*b*). Since the length of the path (circumference of the circle) is $2\pi r$, this condition can be written:

$$2\pi r = n\lambda$$

where n is an integer. For the wavelength λ we can substitute h/p, or h/mv, so that the condition becomes:

$$2\pi r = \frac{nh}{mv}$$

which can be rearranged as

$$mvr = n\frac{h}{2\pi}$$

But *mvr* is the angular momentum of the electron, and so we have in fact obtained the quantization law of angular momentum which is the basis of Bohr's model.

However, this elegant exercise could not help Bohr's model out of the difficulties it had run into, described above. Now, having learnt about the wave nature of the electron, we can ask: are not these difficulties due to the fundamental assumption that the electron moves in a circular path around the nucleus? This path does indeed fit the case of a charged particle orbiting around another charged particle, but it is not necessarily the path that a confined *wave* would choose for itself. And indeed it was found that to arrive at a better model it was necessary to calculate how an actual wave would behave under the circumstances. In other words, a completely new system of laws had to be derived to describe the motion of the small particles, since these particles did not obey Newtonian mechanics. Within a few years such a theory was developed, known as quantum mechanics, and it succeeded in meeting the challenge not only of the electron's motion around the nucleus but of many other problems related to the structure of matter. Unlike the earlier contributions of Planck, Einstein, Bohr and de Broglie, whose ad-hoc models all rested on a limited number of experimental facts, quantum mechanics is a physical theory in the true sense of the word – a comprehensive, complete theory capable of explaining a great variety of phenomena.

In the early development of quantum mechanics there were two different approaches which complemented each other – the 'matrix mechanics' of Heisenberg, and the 'wave mechanics' of Schrödinger. These two approaches are described below.

Heisenberg and the 'magic squares'

In 1925, about 3 years after de Broglie's wave–particle model was proposed, Werner Heisenberg – a young German physicist, aged 24 at the time, lecturing at the University of Copenhagen – challenged the whole idea of formulating models for the structure of the atom. If, argued Heisenberg, our minds are incapable of grasping what is inside the atom – Bohr's electrons or de Broglie's waves, or little goblins, or chess players – let us abandon all attempts to construct any kind of model and stick to quantities which we can measure and understand. Bohr's mistake, said Heisenberg, was that his model rested on quantities which could not possibly be measured, such as the orbit and speed of an electron in an atom. We cannot experimentally determine the location of an electron in an atom without destroying the entire atom. When we deal with a realm which is beyond the reach of experiment we must relinquish all concrete models and representations. Instead, we must restrict ourselves to measurable quantities such as the frequency of light which the atom can absorb and emit – and look for equations to connect these quantities.

It is not easy to follow this advice and construct a world of abstract equations only. The human mind always prefers to think in terms of a concrete model based on images drawn from the perception of the senses – even if it is clear that such a model is only part of the truth. But Heisenberg achieved his goal, in collaboration with his friend, P. Jordan, and their mentor, Max Born (a German-born Jew who emigrated to England when the Nazis seized power. Winner of the Nobel prize for physics in 1954). They arranged the measurable quantities in square arrays of numbers (such arrays are called 'matrices') and by defining mathematical operations between these matrices, they created a consistent quantum mechanical theory. This version of quantum mechanics, which has been known as matrix mechanics, succeeded in explaining certain experimental facts in atomic physics and even predicted unknown phenomena which were verified later experimentally. For his invention of matrix mechanics Heisenberg won the Nobel prize in 1932, when he was only 31 years old.

Another young physicist, Paul A. M. Dirac, born in 1902 in England, pursued the advance of quantum mechanics further by formulating Heisenberg's theory in a more generalized form. Dirac, then a research student at St John's College in Cambridge, received a preprint of Heisenberg's 1925 first paper, and after studying it for 10 days, sat down and wrote an alternative formulation, which presented quantum mechanics as a coherent axiomatic theory. We shall later meet more of Dirac's valuable contributions to modern physics. In 1933 Dirac received the Nobel prize for physics (together with E. Schrödinger).

In the years that followed Heisenberg continued to work on the theory of quantum mechanics and nuclear physics as well as other areas of physics – particularly the theory of ferromagnetic substances, cosmic rays and superconductivity. The number of important contributions he made to physics exceeded that of any other physicist in the twentieth century, except Einstein. During the second world war Heisenberg headed the 'Uranium Project' in Germany, but despite the advanced state of theoretical nuclear physics in Germany the Uranium Project did not achieve practical results. Towards the end of the war Heisenberg was captured by the Americans in the town of Heigerloch close to which he and his assistants were engaged in building a small nuclear reactor. In 1946 Heisenberg returned to Germany to assist in rehabilitating scientific research. He died in Munich in 1976.

Schrödinger and the wave equation

Heisenberg's matrix mechanics and Dirac's formulation of the theory constituted a major revolution in physical theory. But an even greater impact was made by an alternative approach to quantum mechanics presented by Erwin Schrödinger. Viennese-born Schrödinger was deeply impressed by the work of de Broglie and sensed how to pursue further his ideas with the aid of mathematical tools known from the classical theory of waves. Waves, such as electromagnetic and sound waves, obey an equation called the classical wave equation, which relates the spatial variation of the wave to its time variation. This equation enables one to study the properties of various waves. Schrödinger believed that a wave equation should be set up for sub-atomic particles too, which would replace the accepted equations of Newtonian mechanics. In 1926 he produced a series of papers in which appeared for the first time the now-famous 'Schrödinger equation'. This equation soon opened new vistas in atomic physics. (In 1933 Schrödinger won the Nobel prize in physics together with Paul Dirac.)

The Schrödinger equation does not apply to all circumstances since it assumes that the velocities of the material particles involved are much less than the speed of light (and thus does not take into account relativistic phenomena), and that the number of particles does not change (i.e. mass does not vanish or get created). Later more general wave equations were formulated, valid for a wider range of conditions, but nevertheless the Schrödinger equation is a good, simple approximation applicable to very many phenomena, especially at the atomic and molecular level. It explains all the phenomena on which Bohr's model foundered. In principle, it can be used to analyse all chemical reactions.

Application of the Schrödinger equation to the case of an electron wave in an atom – confined in the region around the nucleus due to electrical attraction – yielded a series of solutions, each of which fitted a confined wave of a different energy and angular momentum. Each such solution is called a wave function. Thus the riddle of why the electrons in the atom had fixed and defined energy states was at long last solved. And what was the shape of this wave function of the electron? Did it have the shape of a circular orbit, as Bohr supposed? Not at all! The wave assumes a variety of shapes, creating a sort of symmetrical cloud, such as shown in Fig. 2.7. What is the meaning

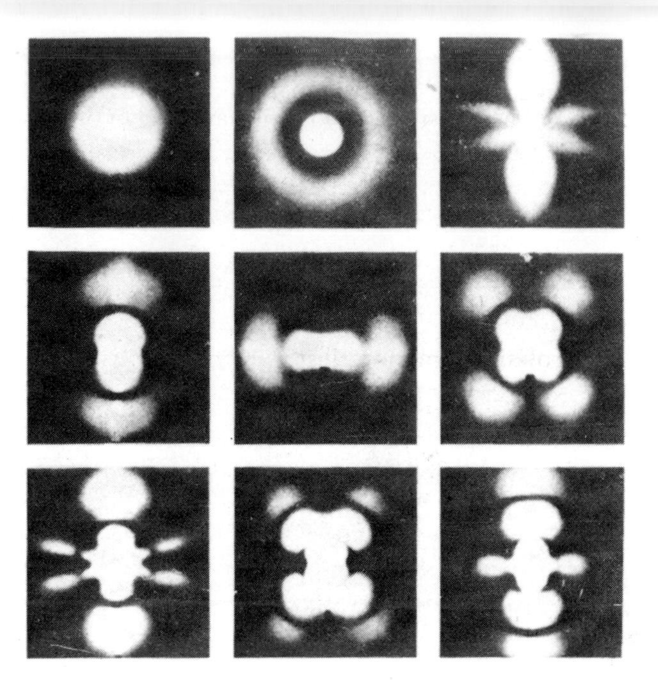

Figure 2.7. Patterns of the electron wave in the atom.

of these patterns? Is every electron a sort of misty cloud that encircles the nucleus? This image, which often appears in popular books on science, is no less misleading than Bohr's orbits. The cloud shapes in the figure are models constructed to demonstrate the solutions of the Schrödinger equation. But such clouds can never be seen in a real atom. If we could take an atom of hydrogen, for example, and perform an experiment to determine the exact position of its electron at a given instant, we would find that the electron was located at a particular defined spot. What then is the meaning of the 'electron cloud'? It turns out that it has a *statistical* meaning. The density of the cloud at a particular point is related to the probability that the electron will be found at that point. Where the cloud is denser the chance of finding the electron there in an experiment is greater.

Along what path within the region defined by the cloud does the electron move? There is no answer to this question – not for lack of knowledge but because of the wave nature of the electron. A wave confined in a box moves throughout its volume, with no definable path. This idea may present some difficulty. After all, we have just said that if we could determine the location of an electron experimentally, we would find that it was located at a well-defined spot. How then can we claim that it is a wave without a path? The answer is that the actual performance of the experiment alters the original character of the atom and causes the electron suddenly to take up a defined position at a particular spot. In other words, the measurement blurs the wave nature of the electron and emphasizes its particle properties.

Another question is: how does the electron have angular momentum if it does not move in a circle around the nucleus? The answer to this is that in quantum mechanics the angular momentum is an independent quantity that is not necessarily connected with a rotational motion, just as momentum is not necessarily connected with mass. Interestingly, it is Bohr's model which supplies a concrete explanation for the angular momentum of the electron in the atom – but fails to explain other properties – while the electron cloud model explains many other properties (for example, chemical bonds between atoms in molecules) but does not supply an explanation for the source of the angular momentum of the electron. Again and again we come up against the realization that no single model can give us a visual picture of all aspects of the atomic world.

Heisenberg's uncertainty principle

In the previous section an experiment to determine the exact location of the electron in the atom was mentioned. But is such an experiment possible at all? Suppose we could overcome the technical difficulties and build up a highly refined instrument capable of doing this; would the capricious dual character of the electron permit such a measurement in principle? Heisenberg, who held that physics should deal only with measurable quantities, was greatly exercised by the question of what could and could not be measured in the atomic world. His computations reached the following interesting conclusion: some of the measurable quantities in the world of small particles are arranged in pairs, and it is impossible to measure the accurate values of both partners at the same time. For example, the momentum and the location of a particle constitute such a pair. Suppose that a particle is moving along a line and we wish to measure its momentum and location along the line at a certain instant. Heisenberg proved that this could not be done with perfect accuracy, no matter how sensitive the instruments. There would always be some uncertainty in the location (Δx) and some uncertainty in the momentum (Δp). If we multiply these two uncertainties, the product will be at least of the order of $h/2\pi$:

$$\Delta p \times \Delta x \approx h/2\pi$$

(Heisenberg's exact formulation was $\Delta p \times \Delta x \geqslant \frac{1}{2}h/2\pi$.) If we succeed in decreasing the uncertainty in the momentum, the uncertainty in location will increase, and vice versa.

This law, known as Heisenberg's uncertainty principle, has been given numerous philosophical interpretations relating to man's inability to reach down to the roots of the laws of nature. In fact, Heisenberg's principle does not imply any limitation on our ability to investigate the sub-atomic world, since we can, in principle, carry out measurements as accurately as we please of each physical quantity at a time. The uncertainty principle is actually a direct outcome of the wave nature of the particles. In order for the wave to have an accurately defined momentum it must contain a relatively large number of wavelengths, and so the location of the particle represented by the wave is not well defined. On the other hand a wave containing a small number of wavelengths is better defined as regards location but less well defined as regards momentum.

In the 1920s, physicists amused themselves devising hypothetical experiments to determine various quantities in the sub-atomic world. For example, they imagined looking at an electron with a microscope sensitive to 'hard' X-rays rather than light rays. In every case it could be proved that Heisenberg's principle held true.

The answer to the question posed above is therefore that it is in principle possible to determine the location of the electron in the atom, but at the cost of a complete blurring of its original momentum. (For example, if an X-ray microscope were used to observe an electron, the high-energy X-ray photon used to make the electron visible would give it so much momentum as to knock it completely out of the atom.)

A similar uncertainty principle relates energy and time. If we perform an experiment to determine the energy E of a particle, at a certain time **t**, then

$$\Delta E \times \Delta t \geqslant \tfrac{1}{2}h/2\pi$$

where ΔE is the uncertainty in the energy, and Δt is the uncertainty in the time of measurement. The more we improve the accuracy in determining the time, the less accurate becomes our knowledge of the energy. In Chapter 8 we shall see that when dealing with short-lived particles we can interpret the last inequality in the following way: ΔE is the uncertainty in the mass of the particle (expressed in energy units) while Δt is its mean life span.

Probability in the quantum theory

The physics of the nineteenth century was based on a deterministic approach, which holds that if we have all the data on a physical system at a given instant we can in principle determine its state at any moment in the future.

This approach seemed well founded. Newton's mechanics had been perfected to a high degree and astronomers could forecast accurately the orbits of the planets and predict eclipses and the appearance of comets. Electrodynamics satisfactorily explained all the known electrical and magnetic phenomena, and thermodynamics made it possible to calculate how much energy would be converted to heat in a given engine, and in what direction a chemical reaction would proceed. Physicists regarded nature as a sophisticated machine operating according to an exact programme from which there could be no deviations.

The advent of quantum mechanics changed all this. It was found that at the atomic and sub-atomic level the future state of a system cannot be predicted by knowing its present state. It is, however, possible to calculate – on the basis of this data – the *probability* that the system will develop in a certain direction. An example of this is the phenomenon of radioactivity. Assume for instance that we have a vessel containing 10^{12} atoms of the radioactive gas radon. Radon is formed when an atom of radium emits an alpha particle. The average lifetime of the radon nucleus is 132 hours. It in turn emits an alpha particle and turns into a polonium nucleus. If we fix our attention on a particular radon atom, we cannot predict exactly when it will disintegrate – it might happen within the next second or a hundred years from now. But we can calculate the *probability* that it will survive for another second or hour or year. We can also predict with confidence that about half the original number of atoms will disintegrate over the next 92 hours (this time is called the 'halflife' of radon). And we can even calculate what the probability is that at the end of the halflife period 49.99 % or 50.01 %, say, of the radon nuclei will remain. (The probability that the number remaining at the end of one halflife is *significantly* different from 50 % is infinitesimal, unless the original number of atoms is small).

The inability to predict the fate of any particular atom does not stem from a lack of data, but from the nature of the process. A *mathematical* analysis of the statistical phenomenon shows that the radioactive nucleus has no 'memory', and the probability that it will decay within the next second does not depend on the time elapsed since it was formed. A radon nucleus newly formed from radium has the same probability of disintegrating within the next second as one which is 200 hours 'old'. This shows that the disintegration is not the result of any hidden internal development but really a random occurrence.

What has been said above holds true for all types of radioactive nuclei, except that the halflife is different for each isotope. Moreover, some nuclides may decay in two distinct modes (by emitting an alpha or a beta particle) and for each of these there is a definite characteristic probability.

The wave nature of atomic systems is closely connected with their probabilistic behaviour. We mentioned above that it is impossible to predict where in an atom an electron will be found when we measure its location, but the *probability* that it will be in a certain spot is determined by the 'intensity' of the wave function at that spot, which can be calculated. It should be noted at this juncture that there *are* measurements whose results are accurately known, for instance the measurement of the electron charge.

The majority of elementary particles are not stable and decay into other particles a certain time after their formation. Neither the lifetime of a single particle nor the products of its decay are fixed, but rather are statistical variables (although there are some particles which always decay into the same products). Also the results of 'scattering processes' in which two particles collide are not fixed. Rutherford developed an expression for the scattering of an alpha particle by a nucleus on the assumption of an electrical interaction between two 'classical' particles. But the quantum mechanical approach holds that even if we know the path of a particular alpha particle and the location of the nucleus, we still cannot calculate the angle of deviation with certainty, but only the probability of various angles of deviation (for instance, by means of the Schrödinger equation). It turns out, however, that the expression obtained for the scattering of a large number of alpha particles is identical to Rutherford's expression.

Conceptual revolution

The probabilistic interpretation of quantum mechanics and its relation to wave functions, first proposed by Max Born, were soon adopted by the majority of physicists. By doing this, modern physics actually gave up the principle of determinism, or absolute causality, which had been an implicit assumption of classical physics. According to this principle, each given physical state has a single, well-defined, consequence.

Curiously, Einstein was among the few to whom the renunciation of determinism seemed repugnant. He could not accept the statistical basis of quantum mechanics and on one occasion he wrote: 'Quantum mechanics is very impressive. But an inner voice tells me that it is not yet the real thing. The theory produces a good deal but hardly brings us closer to the secret of the Old One. I am at all events convinced that *He* does not play dice' (from a letter to M. Born, dated 4 December, 1926). As late as 1953 Einstein still claimed that founding the physical theory on this approach was, in his opinion, very unsatisfactory. However, all attempts to develop an alternative theory, which will treat atomic phenomena without introducing probability into the basic laws of nature, were doomed to fail. Quantum mechanics is today largely accepted as a true description of the atomic world.

Einstein's objection, in spite of being a minority opinion, reflected the perplexity and confusion which were the lot of many physicists in the 1920s and 1930s. Theirs was a transition generation, which had witnessed the collapse of the old concepts and the budding of the new physics. They were awarded with the unique elation experienced by pioneers in an unknown land, but at the same time underwent a mental shock, while confronting the new revolutionary ideas such as the wave–particle duality, the uncertainty principle and abolition of the deterministic basis of physics. To mention one example: when O. Stern and M. v. Laue, who were later to contribute significantly to modern physics, first heard about Bohr's model, they declared that if by any chance it should prove correct, they would abandon physics. For these physicists, whose education had been rooted in the previous century, the assimilation of the quantum notions involved a deep conceptual crisis. This crisis had a profound effect on the scientific thought in general. It admitted into science an element which has been a cornerstone in philosophy – doubt. For the first time in the history of physics, the physicists had to re-examine everything they knew, and to become accustomed to the idea that every theory, firm and stout as it may be, must be perpetually tested by new experiments and adapted, when necessary, to the updated results.

It should be noted that the success of quantum mechanics did not render classical mechanics (for which material bodies are objects of well-defined dimensions rather than waves) into a wrong or obsolete theory. Classical, or Newtonian, mechanics is still the 'true' theory in most cases when dealing with systems much larger in dimension than atoms and molecules. Indeed, applying quantum mechanics to such systems usually yields the same results as does classical physics. (The same is true when one applies special relativity to a motion at a low velocity. The results coincide – within the practical accuracy limits – with those provided by non-relativistic equations.)

At the end of section 2.4 we realized that the wavelike nature of a 'large' body is indiscernible, since its de Broglie wavelength is so tiny. A similar effect emerges when the uncertainty principle is applied to macroscopic ('large') systems. If we wish, for instance, to determine the velocity and location of a moving car at a certain instant (e.g. by a motion picture camera), we shall find that the uncertainty associated with Heisenberg's principle is negligibly small compared with the uncertainty imposed by the limitations of our measuring devices. (This explains why Heisenberg's principle was not discovered experimentally prior to the advent of quantum mechanics.) The fact that the quantum mechanical predictions concerning macroscopic systems usually coincide with those of classical mechanics was formulated by Bohr as 'the correspondence principle'. It should be noted, however, that some macroscopic systems do require quantum mechanical treatment. A neutron star and a superconductor represent two examples.

2.6 Properties that characterize the particle – the discovery of spin

A suit made by a tailor has an infinite number of variations – in type and colour of cloth, style and quality of execution. Even in mass production it is difficult to get two absolutely identical suits. The situation is different in the case of elementary particles. Their elementary, fundamental nature is expressed in the fact that they are characterized by a small number of basic properties such as mass, electric charge, energy and momentum. Two electrons of the same energy are absolutely identical and we cannot distinguish between them in any way at all.

In 1932 four elementary particles were known to physicists: the photon, electron, proton and neutron. (In the same year an additional particle – the positron – was discovered. We shall meet this newcomer in the next chapter.) The photon is characterized by a fixed speed and zero mass, and its energy determines its momentum and wavelength. The other three particles can have any speed between zero and the speed of light (but cannot actually equal the speed of light) and the speed of each particle determines its momentum, energy and the wavelength of its associated wave. The rest mass and the charge are fixed properties of the particle and do not change.

Does this list cover all the properties of the particle? Not exactly. The neutron and proton, for example, differ not only in electric charge but also in stability: the proton, like the electron, is stable, at least while dealing with time intervals which are not exceedingly long, whereas the free neutron outside the atomic nucleus decays spontaneously. Also the interactions of the particles differ: a collision between an electron and proton will end up differently from that between a neutron and a proton. Physicists strive to reach a theory which will show how *all* the properties of elementary particles spring from a small number of simple basic properties, like electric charge which can be represented by a small integer: 0, 1 or -1 (in units of the electron charge). Over the years more and more properties have been discovered which can be represented by simple small numbers, known as 'quantum numbers'. One of these properties is the spin.

What is spin?

In 1925 two young students from the University of Leiden in Holland – S. Goudsmit and G. Uhlenbeck – put forward a startling idea: an electron, in addition to its motion round the nucleus (orbital motion) also revolves on its own axis, just as the earth orbits the sun and at the same time revolves on its own axis. This rotation of the electron – which they called 'spin' – occurs at a fixed, unchanging angular speed, even when the electron is outside the atom, and is totally independent of the linear speed or environment of the electron. The magnitude of the angular momentum associated with this spin is $\frac{1}{2}h/2\pi$. It will be recalled that the angular momentum of the orbital motion according to Bohr's model is $nh/2\pi$ where n is an integral number.

Fig. 2.8 illustrates (not to scale) the hydrogen atom according to Bohr's model, with the addition of the spin. The spin of the electron is conveniently indicated by an arrow protruding from it, representing its *axis* of rotation. This arrow also indicates the *direction* of rotation, in the following manner. Let us imagine that the arrow is replaced by an ordinary ('right-handed') screw. The direction in which the arrow points has the same relation to the electron rotation as the direction of advance of the screw has relative to the direction of the turn of the screw (see Fig. 2.9). Henceforth we shall talk of *direction of spin*, meaning the direction of the arrow representing the axis of rotation.

If the electron were simply a revolving sphere, that would be the end of the story. But since it is electrically charged, and an elementary particle subject to the laws of quantum mechanics, the spin has two important consequences. First: a revolving electric charge acts as a magnet, and therefore the electron is in effect a small magnet. Second: due to the properties of angular momentum in quantum mechanics, the spin of the electron has exactly two possible orientations with respect to any defined axis in space. If we perform any measurement whatsoever to determine the angle between the direction of the electron spin and any given direction in space, we find

Samuel A. Goudsmit and George E. Uhlenbeck were in their mid-twenties when they stumbled across the notion of spin. They discussed it with their Professor, P. Ehrenfest, who was impressed by the idea, though not blind to the problems associated with attributing a rotational motion to the electron. He asked the two to deliver him a short paper about their model, and then sent them to consult H. A. Lorentz who was the esteemed authority among Dutch theoretical physicists. Lorentz studied their idea carefully for several days, and then showed them that it contradicted all the known data about the electron mass and dimensions. (The apparent contradiction arose, of course, because the spin was treated as an actual rotation.) The two are said to have wanted to withdraw their paper but it was too late – Ehrenfest had already sent it for publication. When published, the model encountered some objection, but eventually was proved to be correct.

A 19-year-old physicist, R. Kronig, arrived at the idea of spin simultaneously with Goudsmit and Uhlenbeck. He was

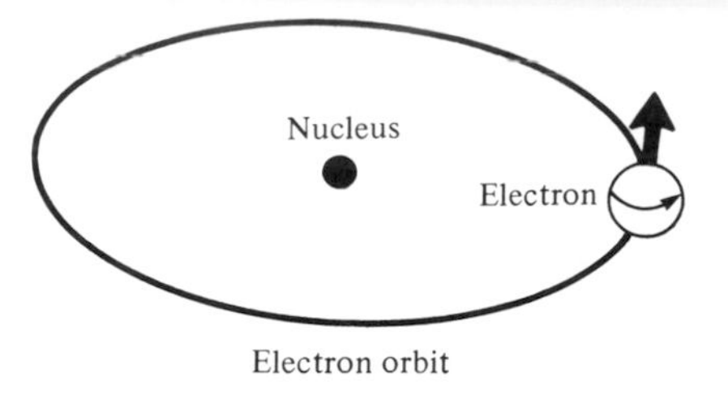

Figure 2.8. The electron spin and orbit in Bohr's model.

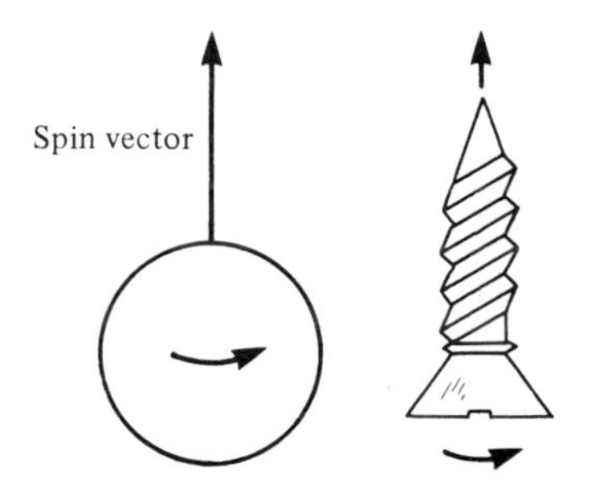

Figure 2.9. The spin vector is related to the 'rotation of the electron around its axis' as the direction of advance of a 'right-handed' screw is to the trend of its rotation.

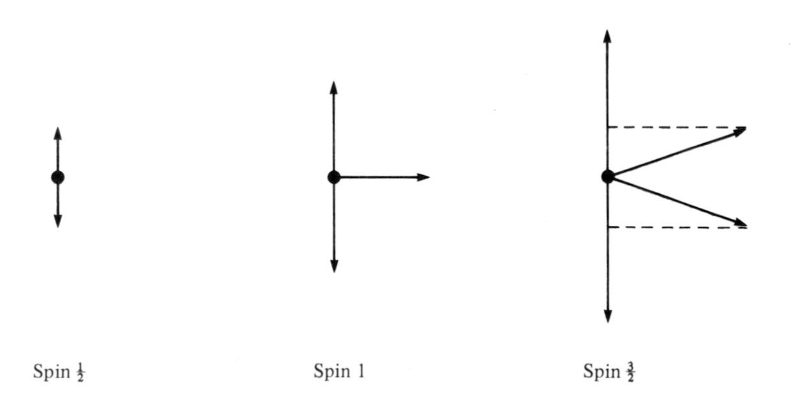

Figure 2.10. The projection of the spin on a defined axis can take only certain discrete values, in steps of $h/2\pi$. A particle of spin $\frac{1}{2}$ has two possible orientations relative to any defined direction in space, spin 1 particles have three allowed orientations and spin $\frac{3}{2}$ particles – four.

visiting Pauli and discussed the hypothesis with him. Pauli pointed out the difficulties and convinced Kronig that the idea had no foundation. When the article of Goudsmit and Uhlenbeck appeared, Kronig thereby published a refutation of the idea, giving Pauli's arguments. However, the model was eventually proved to be correct.

In 1927 Goudsmit and Uhlenbeck moved to the USA. Despite the recommendations of various physicists, Uhlenbeck and Goudsmit never got the Nobel prize, perhaps because they had ignored the difficulties of ascribing a rotational motion to the electron, and let others cope with them. In 1979, however, Uhlenbeck (then at Rockefeller University in New York) was awarded the Wolf prize in Israel. Goudsmit had died several months earlier.

that the angle is *always* either 0° or 180° – in other words, the spin is either parallel or antiparallel to the chosen direction. For example, if an electron beam is acted on by a weak magnetic field which is not strong enough to force all the spins to align themselves in the same direction, then some of the spins will arrange themselves in the direction of the magnetic field and the rest in exactly the opposite direction. (A group of small 'classical' magnets in the same circumstances would be arranged in random directions relative to the field.) The reason for this phenomenon is that not only is the angular momentum quantized (i.e. appears as a multiple of the fundamental quantity $h/2\pi$) but also its projection on a defined axis (see Fig. 2.10). A particle whose spin equals 1 (measured in units of $h/2\pi$) can take up three possible positions. In general, calculation shows that for a spin equal to s (in units of $h/2\pi$) there are $2s + 1$ possible positions.

How did Goudsmit and Uhlenbeck stumble upon the idea of spin? Was it through analogy between the atom and the solar system? Well, by 1925 physicists were wiser than to compare the atom with macroscopic systems like the solar system. Their suggestion was based on some more solid experimental evidence. Firstly, close analysis of the spectrum of light emitted by atoms showed that sometimes a spectral line was in fact composed of two very close lines (a doublet). This meant that there could be pairs of very close electron energy states. The concept of spin with its two possible orientations accounted for this phenomenon very well. It also accounted for the influence of an external magnetic field on the spectrum of various atoms. Moreover, during the years 1921 to 1924 two physicists in Hamburg – O. Stern (1943 Nobel prize laureate who emigrated to the USA in 1933) and W. Gerlach – performed a series of experiments which pointed clearly to the existence of spin.

Stern and Gerlach wanted to study the effect of a magnetic field on atoms. For this purpose they produced a highly non-uniform field by means of a magnet with specially designed poles – one of the poles being tapered to a sharp point (see Fig. 2.11). The lines of force converge on the point, and the field gets stronger and stronger in the upward direction. Suppose that a small bar magnet finds itself in this field with its north pole pointing up. The downward force on its lower end will be weaker than the upward force on its upper end, and so there will be a resultant force acting on it in the upward direction. A magnet whose north pole points downwards will experience a resultant downward force of similar magnitude. A magnet making an angle of between 0° and 180° with the vertical direction will be subject to a smaller force than a magnet parallel to

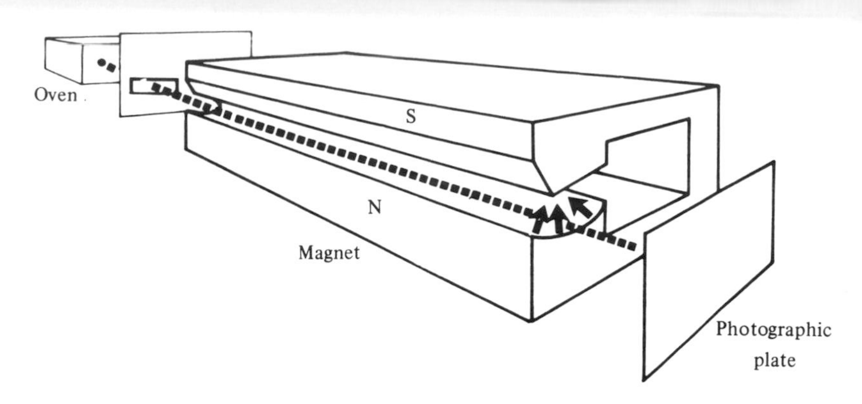

Figure 2.11. The Stern-Gerlach experiment. A beam of silver atoms ejected from the oven passes between poles of a magnet. The magnetic field becomes stronger in the upward direction.

the vertical. What will happen if a beam of particles, each of which is a small magnet pointing in a random direction, is drawn into this field? Each little magnet will experience a different resultant force according to its angle of inclination, and we would expect the beam to spread across the whole horizontal axis. Stern and Gerlach investigated beams of atoms of various types, and found that the beams did break up – not into a continuous wide beam but into a small number of narrow beams. For example, a beam of silver atoms broke up into two beams, which were diverted upwards and downwards. The result was explained in the following way. Both the orbital motion of the electrons and their spins contribute to the magnetism of the atom (an additional contribution from the nucleus is negligibly small). Out of the 47 electrons in the silver atom, 46 form a spherical structure, the magnetism of which is null. The magnetism of the silver atom thus springs solely from the spin of the 47th 'external' electron. The splitting of the beam into two parts proves that the spin of this electron can take up only two orientations regarding a defined direction in space.

Is the electron really revolving on its axis?

At first glance it appears that all the phenomena related to the electron spin – angular momentum and magnetism – are well accounted for by assuming that the electron rotates on its axis. But a closer look shows that it is difficult to interpret the spin in such classical terms. If we assume that the electron is a charged sphere of a certain diameter, we may find that in order for the angular momentum and magnetism to have the required values its surface would have to move at a speed greater than that of light, which is impossible. Today spin is no longer interpreted as resulting from rotary motion, just as the angular momentum of orbital electrons in the atom is not regarded as resulting from circular movement around the nucleus. So it is more accurate to say that the electron has an intrinsic angular momentum of $\frac{1}{2}h/2\pi$, called spin, *as if* it were rotating about its axis.

In 1925 Pauli showed how the Schrödinger equation could be modified to take into account electron spin and to calculate, for example, the probability that the spin would point in a given direction. Pauli's method achieved great success but it did not shed light on the origin of the spin. Moreover, the important properties of the electron spin – such as the fact that it has two possible states relative to any defined direction and that it is associated with magnetism – did not emerge directly from Pauli's calculations, and had to be accepted as given.

A significant advance came in 1927 when the British physicist Paul Dirac proposed a relativistic wave equation for quantum systems. Dirac's aim had been to derive a wave equation similar to Schrödinger's which would take into account Einstein's relativity theory and would therefore be applicable to speeds close to the speed of light. Not only did he succeed in this, but as a special bonus it turned out that his equation also predicted – on the basis of the most general principles and without any pre-assumption – all the electron properties resulting from its spin. (Further successes of Dirac's equation are described later.)

Spin and magnetic moment

During the years it was found that not only the electron possesses a spin, but other elementary particles do as well. Spin is a fundamental property of a particle, along with its electric charge and mass. The spins of all the particles known at present are integral multiples of the 'natural unit' $\frac{1}{2}h/2\pi$. In other words spin can take the values 0, $\frac{1}{2}$, 1, $\frac{3}{2}$, etc. in $h/2\pi$ units. Experiments showed the spin of the proton and neutron to be $\frac{1}{2}$ and that of the photon to be 1.

We have mentioned the fact that because of their spin, the electron and other particles behave like small magnets. The magnetism of a bar magnet is measured by its 'magnetic moment', which is defined by the energy needed to turn the magnet from a position parallel to a given magnetic field to a position perpendicular to the field (Fig. 2.12). In the metre–kilogram–second system of physical units, the strength of the magnetic field is measured in tesla.* (To

* Named after Nikola Tesla who invented the first alternating current electric motor in 1883.

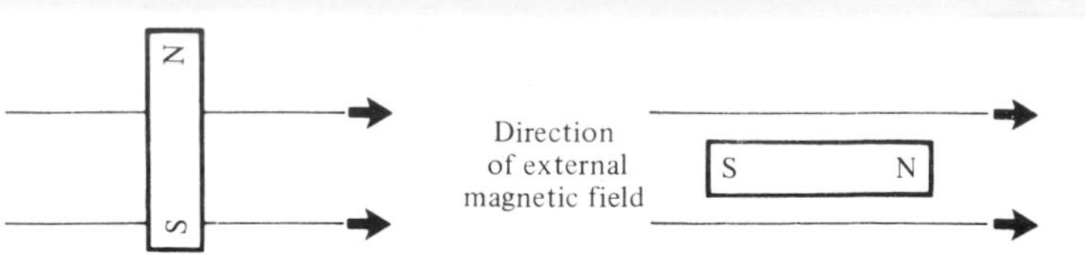

Figure 2.12. The magnetic moment of a bar magnet is defined by the energy required to turn the magnet from a position parallel to an external magnetic field (whose direction is indicated by arrows) to a position perpendicular to it.

give an idea of this unit, the magnetic field strength of the earth close to its surface is between 0.25×10^{-4} and 0.5×10^{-4} tesla, varying within this range from place to place. Between the poles of a toy magnet there is a field of several hundredths of a tesla, and in a big electromagnet the field strength can reach several tesla.) The unit for measuring magnetic moment is joules per tesla (the joule being a unit of energy). If one joule of energy is needed to turn a magnet from a position parallel to a magnetic field to a position perpendicular to it, when the strength of that magnetic field is one tesla, then the magnetic moment of the magnet is one joule per tesla.

The magnetic moment of the electron is conveniently measured in a unit called the Bohr magneton, denoted μ_e. This unit equals 9.27×10^{-24} joules per tesla (this, according to Bohr's model, is the magnetic moment of the hydrogen atom when the electron is in the ground state).

The magnetic moment of the electron was measured to a very great accuracy (up to 10 digits) and was found to be 1.001 159 652 μ_e. Neither classical physics nor quantum mechanics could explain how the electron spin induces such a value for the magnetic moment. One of the successes of Dirac's 1928 equation had been its reproduction of the value of the Bohr magneton from first principles, instead of Bohr's heuristic derivation. However, more precise measurements had since yielded a slightly larger value. In 1948, a new and more sophisticated theory provided the means of including in the calculations the additional effects due to virtual processes, i.e. processes that can take place beneath the threshold of observability set by the Principle of Uncertainty; for example, the emission and subsequent reabsorption of photons, which is going on all the time, making up what is usually called a 'photon cloud' surrounding the electron in the quantum world. How impressive, when the new theory gave the value of the electron's magnetic moment to the tenth digit, and this with a perfect fit with experiment! The theory, called quantum electrodynamics (QED for short), was put forward by three physicists working independently: the Japanese Sin-itiro (meaning 'first-born') Tomonaga, and the Americans Richard Feynman and Julian Schwinger. The theory aimed at presenting the proper quantum treatment for electromagnetic fields and electric charges, and was based on an idea proposed by Dirac in 1927: to treat the electromagnetic field itself not as a continuous quantity but as something composed of discrete entities or quanta. At the same time, the electron wave function is also treated as a field. The theory managed to overcome some mathematical difficulties that withheld the development of the basic idea for some twenty years, and was very successful in predicting experimental results. Even today it is considered to be the most precise existing scientific theory. It is interesting that Dirac himself regarded the new theory with suspicion, just as Einstein had received quantum mechanics.

Actually, the clue to the QED theory was concealed in papers published by E. C. G. Stueckelberg in 1934–38, and more explicitly in his paper of 1947. Unfortunately his formulations were obscure and difficult to use. Had the theorists been able to follow them, the magnetic moment of the electron and other results of QED could have been calculated much earlier.

Feynman, Schwinger and Tomonaga shared the 1965 Nobel prize in physics. They could possibly have been awarded that prize several years earlier, but Niels Bohr also suspected the new theory and his negative attitude deterred the Nobel Committee from acknowledging it.

The decision to give the Nobel prize to the authors of QED was made only after Bohr's death. When it was published, R. Oppenheimer ('the father of the atomic bomb') was in Israel, as Yuval Ne'eman's guest. He sent Feynman a telegram from Tel Aviv with only one word: '*Enfin*' ('at last', in French).

The photon has no magnetic moment. A particle of zero mass can apparently have neither a charge nor a magnetic moment. In 1933 Stern improved the Stern–Gerlach experiment and measured the magnetic moments of several nuclei. The magnetic moment of the proton was found to be smaller by three orders of magnitude than that of the electron but anomalous in the sense that it was almost three times larger than expected, if we replace the electron mass by the proton mass in the formula for μ_e. The physicists were surprised to find in 1936 that the neutron, despite its zero charge, had a magnetic moment too! It was of the same order of magnitude as that of the proton, and its direction was as if the neutron carried a negative electric charge. These facts hinted that the neutron (and also the proton) have an internal structure which includes positive and negative charges, because magnetism always involves the motion of charges.

The Pauli exclusion principle

Spin is related to another property which on the face of it has nothing to do with angular momentum. Physicists found that the state of the electron in an atom could be characterized by four quantum numbers: the first (n) indicates the 'shell' or energy state which the electron occupies; the second (l) determines the orbital angular momentum of the electron; the third (m) determines the inclination which the direction of the orbital angular momentum will take up relative to a defined direction (e.g. a magnetic field); the fourth (m_s) determines the direction of the spin (for which there are only two possibilities). A basic law of atomic physics says that no two electrons in an atom can have identical sets of quantum numbers (note that this means that they can be distinguished from each other). In the helium atom, for example, there are two electrons and they are identical in three of their quantum numbers, but opposite in their spins. The state of the electrons in every atom can be accounted for with the help of this law. The principle was formulated in 1924 by the Austrian Wolfgang Pauli (who contributed greatly to quantum mechanics) and earned him the 1945 Nobel prize in physics. It is called the Pauli exclusion principle or Pauli's principle for short.

In due course it has been found and proven theoretically that the Pauli exclusion principle is valid for any particle whose spin is not integral, i.e. is $\frac{1}{2}$, $\frac{3}{2}$, $\frac{5}{2}$, etc. The laws of behaviour of these particles are embodied in 'Fermi–Dirac statistics', named after the physicists who formulated them. These are the statistics characterizing distinguishable objects. The particles themselves are called *fermions*. When several fermions *of the same type* are close to each other (in other words, when their wave functions overlap) they cannot be in identical states; thus they must have non-identical sets of quantum numbers. Protons and neutrons are also fermions (spin $\frac{1}{2}$), and thus in a nucleus they populate different energy levels just as the electrons in the atom do.

Particles with integral spin (0, 1, 2, etc.) are called *bosons*, after the Indian physicist S. N. Bose who (together with Einstein) investigated them. They do not obey the Pauli exclusion principle. Thus an unlimited number of bosons with identical quantum numbers (e.g. photons of identical energy) can be concentrated in a particular region of space. Bosons are thus indistinguishable. The laws of behaviour of bosons are called 'Bose–Einstein statistics', and are based on the statistics of indistinguishable objects. It can be shown that the difference between fermions and bosons is related to the connection between the spin and the symmetry of the wave function of the particles.

The size of a particle

Before leaving the subject of the properties of particles, we shall touch on another interesting question: how large are the particles? In the case of a photon, which has no mass or charge, it may be assumed that the energy it carries is located in a region of space whose dimensions are of the order of a wavelength (although under those circumstances where the photon displays its particle nature its energy seems to be concentrated at a point). Neutrons, protons and electrons have mass. Does this mass fill a region of space whose diameter can be measured? In the case of the electron, it was difficult to answer this question since – owing to the uncertainty principle – the electron is never at rest. An electron in an atom – even in the lowest orbit – is 'spread' over a region whose diameter is of the order of the atomic diameter. But in phenomena where the electron displays a particle character (e.g. in the Compton effect) it appears to have a much smaller diameter.

In the 1930s, physicists devoted much attention to the question of the electron dimensions and tried to develop a theoretical formula for its diameter. They arrived at the formula $r = e^2/mc^2$ where e and m are the charge and mass of the electron and c the speed of light. This gives a value of 3×10^{-13} centimetres for the radius. In some phenomena – such as the scattering of light by electrons in the atom – the observations are consistent with this radius, but according to more recent experimental data and the theoretical framework provided by QED, the electron is practically a point particle whose mass and charge are concentrated in a region smaller than 10^{-16} centimetres.

As far as protons and neutrons are concerned, experiments show that their mass is concentrated in a region with a diameter of about 1.2×10^{-13} centimetres. In experiments on the scattering of particles by nuclei, for example, it is found that the volume of a nucleus containing A nucleons equals the volume of A spheres of radius 1.2×10^{-13} centimetres glued to each other.

Chapter Three

The discoveries of the 1930s and 1940s

In 1932, with the discovery of the neutron, it seemed that the picture of the atomic world was complete. This picture featured four elementary particles (photons, electrons, protons and neutrons), the properties of which are summarized in Table 3.1. The atomic nucleus was composed of neutrons and protons, and the behaviour of the electrons surrounding the nucleus was well explained by quantum mechanics (which successfully explained many other phenomena in the atomic and sub-atomic world). Actually, this picture was not a perfect one. There were several unsolved problems such as the process of beta decay and the nature of the force holding the nuclear components together. The theories which attempted to solve these problems predicted the existence of additional particles, and the experimentalists searched for and discovered them – along with other particles that the theoreticians had not predicted. The particle research of the 1930s and 1940s is conveniently described under four headings:

(a) discovery of the positron and understanding the role of anti-particles;
(b) the neutrino and the 'weak force';
(c) Yukawa's theory of the 'strong force';
(d) discovery of the muon and pion.

The experimental research of the elementary particles in that period depended quite heavily on cosmic rays, which are briefly described in the next section.

3.1 Cosmic rays

At the beginning of the twentieth century, physicists studying radioactivity observed that the leaves of an electroscope discharged

Table 3.1. *Particles discovered up to 1932*

Particle	Mass kg	Mass MeV	Electric charge (*e* units)	Spin ($h/2\pi$ units)	Magnetic moment (joules per tesla)
Electron	9.109×10^{-31}	0.511	-1	$\frac{1}{2}$	-9.27×10^{-24}
Proton	1.673×10^{-27}	938.26	1	$\frac{1}{2}$	1.4×10^{-26}
Neutron	1.675×10^{-27}	939.55	0	$\frac{1}{2}$	-0.96×10^{-26}
Photon	0	0	0	1	0

and collapsed slowly even with no radiation source in the vicinity. It became clear that this phenomenon was caused by a continuous flux of low-intensity radiation which must exist everywhere on the earth's surface. At first it was thought that this radiation originated from the natural radioactive elements in the earth's crust, but in 1910 the Austrian physicist Victor Hess went up in a balloon and found that the radiation grew stronger the higher up he went. This and other experiments proved that the radiation originated outside the earth's atmosphere. It was given the name 'cosmic rays', a name proposed by the American physicist Robert Millikan who played an important part in investigating it. (Hess, however, is regarded as the discoverer of cosmic rays and he shared the 1936 Nobel prize in physics for this discovery.)

A large number of studies, carried out with instruments taken to the upper atmosphere in rockets and balloons, showed that cosmic rays are composed of fast particles, and that the particles reaching the earth's surface are not the original ones entering the atmosphere. The 'primary' cosmic particles coming from outer space are nuclei of various elements stripped of all their electrons. Most of them are nuclei of hydrogen atoms – single protons – but there are some heavier nuclei (cosmic radiation provides one of the proofs that 99 % of the matter in the universe is hydrogen and helium). The primary particles reach the atmosphere at very high speeds – the slowest among them have energies of 10^9 electronvolts and the fastest have fantastic energies of up to 10^{19} electronvolts. When they reach the upper atmosphere they collide with the nuclei of atoms in the air and the result is a shower of secondary particles (see Fig. 3.1).

What is the source of the 'primary' particles? Some of the slower ones come from the sun. The origin of the others is still unclear. They may be traces of a supernova – a giant explosion of a star – or they may come from some enormous unknown sources of energy somewhere outside the milky way galaxy. However, the processes

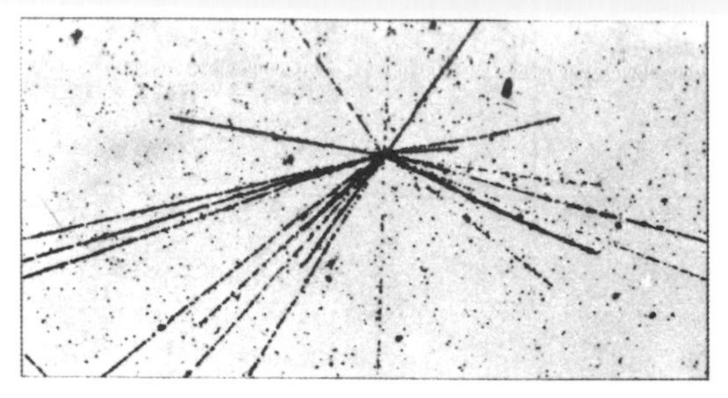

Figure 3.1. The tracks of particles produced in the collision of a fast-moving primary cosmic particle with a photographic plate.

occurring when these energetic particles reach the atmosphere provided exciting information for the elementary particle physicists in the 1930s and 1940s, as we shall see.

3.2 Discovery of the positron

In 1928 Dirac published his relativistic wave equation mentioned above (p. 57), from which there emerged, among other things, the property of the electron that we described as 'spin'. However, it was soon found that the equation provided for the existence of *four* different states of the electron and not the observed two (spin up and down). The two additional states described an electron with negative energy, and at first the physical meaning of this result was not clear. In the end Dirac reached the daring conclusion that there must exist a particle similar to the electron but having a positive charge.

In 1932 Carl Anderson, working at Caltech (the California Institute of Technology), Pasadena, discovered a positively charged particle with the same mass as the electron among cosmic ray particles. The particle was discovered with the aid of a cloud chamber in which a magnetic field was applied. The magnetic field causes the path of a charged particle to curve, the curvature of a positively charged particle being opposite to that of a negatively charged one. To establish the direction of travel of the particle, Anderson placed a lead plate in the chamber and studied the photographs of those particles that had passed through the plate. Since the particle lost some of its kinetic energy in the plate, its speed decreased and the curvature of the path increased after the plate had been traversed. So the direction of travel could be determined, and since the magnetic field direction was also known it could be easily established whether the charge of the particle was positive or negative. In cloud chamber pictures such as that shown

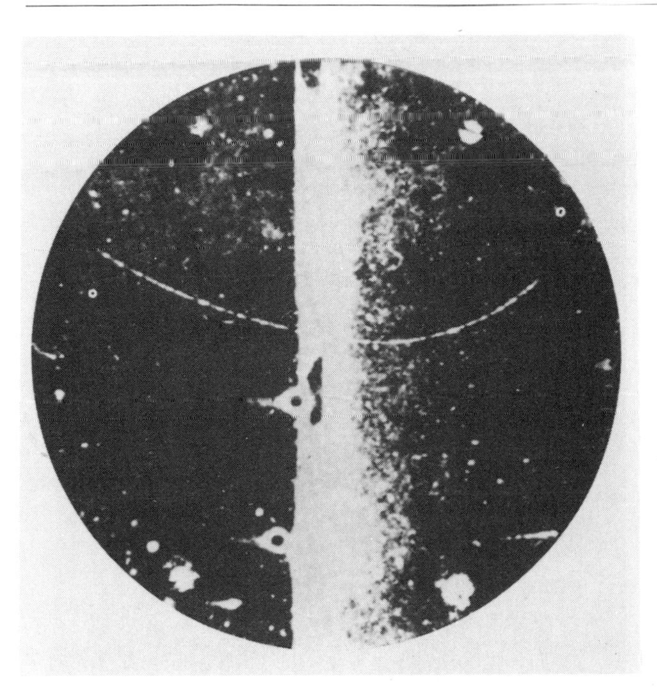

Figure 3.2. The path of a positron of the secondary cosmic rays, as photographed in a cloud chamber. The curvature of the path is larger to the right of the lead plate than to its left, indicating that the particle travelled from left to right. Knowing that the direction of the magnetic field was into the page, it could be determined with certainty that it was a positively charged particle. (From C. D. Anderson, *Physical Review,* **43** (1933) 491.)

in Fig. 3.2, the presence of 'positive electrons' in secondary cosmic rays was clearly demonstrated. The ratio e/m showed that the mass was the same as that of the electron. The particles were named 'positrons' and denoted e^+. (Anderson took part in other discoveries related to cosmic rays, and in 1936 he shared the Nobel prize in physics with V. Hess.)

It was later found that certain artificial radioactive isotopes emit positrons, for example an isotope of phosphorus:

$$_{15}P^{30} \rightarrow {}_{14}Si^{30} + e^+$$

In this process, which is called β^+ emission, a proton in the nucleus transforms into a neutron and a positron (and another particle, as we shall soon see).

According to Dirac's theory, a photon of energy greater than 1.022 MeV (double the rest mass of the electron) could turn into an electron–positron pair, and the phenomenon was indeed observed in cosmic rays. This process cannot occur in a vacuum, since it cannot then obey both the conservation of energy and momentum. (For example, suppose that the photon energy is exactly 1.022 MeV. If it turns into an electron–positron pair the particles would have to be at rest, since all the photon energy would be converted to mass, leaving nothing for kinetic energy. Thus, the particles would have zero momentum, meaning that all the original momentum of the photon had vanished – which is contrary to the law of conservation of momentum.*) Thus the process – called 'pair production' – can only take place in matter in the vicinity of a nucleus which can take up the excess momentum of the photon. This is one of the processes by which energetic gamma rays lose energy in passing through matter. The positron formed in this way does not survive long. When it collides with an electron a process known as 'annihilation' occurs, the two particles disappearing and transforming into a number of photons:

The first cloud chamber photograph of a 'shower' of electrons and positrons produced by cosmic rays was taken at the Cavendish Laboratory at Cambridge by P. M. S. Blackett and the Italian G. Occhialini. It was said that when Occhialini saw the developed photograph he ran to Rutherford's home to tell him the news, and in his excitement he kissed the maid that opened the door.

$$e^+ + e^- \rightarrow n\gamma$$

Pair production and annihilation are impressive illustrations of the equivalence of mass and energy, and of the fact that the difference between a particle with mass and one without mass is not so crucial after all.

Particles and anti-particles

According to Dirac's equation we should expect that not only for the electron, but for every other charged particle there exists a 'mirror particle' identical to it but with an opposite charge. Indeed in 1955 the anti-proton was discovered in a particle accelerator (see section 4.2). It was later actually found that all other charged particles indeed do have their 'anti-particles'. This name and the term 'anti-matter' are somewhat misleading, since the anti-particles are true particles in every respect, with the same mass as their particle partners. Each anti-particle is identical to its corresponding particle not only in mass and spin but also in lifetime, and it is denoted by the same letter with a bar on top. For example, the anti-proton is written $\bar{p}$ or $\bar{p}^-$. The positron ('anti-electron') is an exception – it is denoted just e^+.

We have said that there are anti-particles for all charged particles. But some uncharged particles – such as the neutron – have anti-particles as well. Since neutrons and anti-neutrons are uncharged they cannot be distinguished by applying an electric field but, as we shall see, there are additional 'charges' besides the electric charge that

* Conservation laws are discussed in Chapter 7.

characterize particles. These 'charges', known by names such as 'strangeness' and 'baryon number', cannot always be detected by applying any kind of force field to the particle, but they obey certain conservation laws (similar to that for electric charge). The 'charges' of the neutron and anti-neutron have opposite signs, and this expresses itself in processes where particles decay or are created. Moreover, in the anti-neutron the magnetic moment is in the direction of the spin whereas in the neutron it is in the opposite direction.

A good illustration of the difference between a particle and an anti-particle is in the process of annihilation. The reaction between two colliding neutrons is totally different from that between a neutron and anti-neutron. In the latter case the two particles disappear just like a positron–electron pair, except that instead of photons, particles called mesons are generally formed.

All particles have anti-particles, except those which are their own anti-particles. The photon is an example of a particle which is its own anti-particle, and we shall meet others later. Naturally in all these cases the electric charge and all other charges must be zero.

We have said that particles and anti-particles are identical in all properties (except for their charge) and are equally stable. Why then are the electron and proton such common particles and the positron and anti-proton so rare? The answer lies in the phenomenon of annihilation. A world in which electrons and positrons are equally common cannot survive long and will explode within a short time by annihilation processes. An atom consisting of a nucleus made up of anti-protons and anti-neutrons, surrounded by positrons, will be as stable as an ordinary atom – so long as it does not meet an ordinary atom. Such a meeting would be fateful for both atoms – all their mass would be converted into energy. Luckily the universe (or at least the galaxies in our region) is composed of particles and contains almost no anti-particles.

Indeed it was once believed that in other regions of the universe anti-matter dominates, and that some gigantic releases of energy observed in outer space might result from the meeting between matter and anti-matter. Today, however, it is believed that the universe did start with equal amounts of matter and anti-matter, but physical processes in its early evolution violated the balance and caused the net surplus of matter.

3.3 The neutrino and the weak force

One of the riddles facing physicists during the 1920s was the phenomenon of beta decay. Their research into the properties of alpha decay did not present any particular difficulties. They found that all the alpha particles emitted from a given nuclide may possess the same kinetic energy; alternatively, they may each possess one of a number of specific energies. When the energy of the particle is less than the maximal energy of alpha particles emitted from a given nuclide, a gamma photon possessing the surplus energy is also emitted. Calculations show that the energy liberated in this process

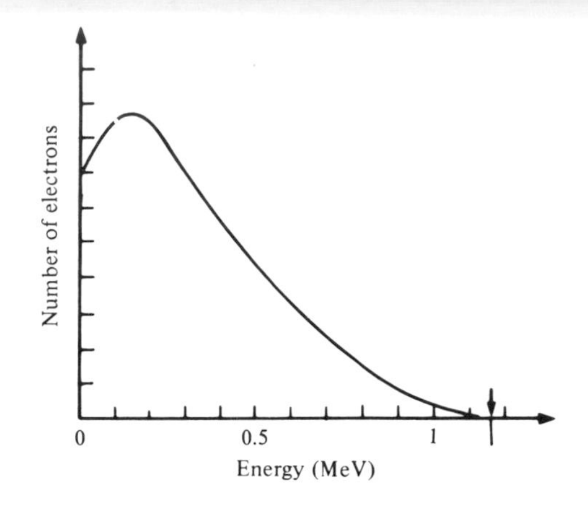

Figure 3.3. The energy distribution of the electrons emitted in the beta decay of bismuth 210. The kinetic energy of these electrons is between zero and 1.17 MeV.

is precisely equal to the difference between the mass of the original nucleus and the mass of the products (the alpha particle and the remaining nucleus). In other words – energy is conserved in this process. Alpha decay was found to obey the laws of conservation of momentum and angular momentum as well.

On the other hand, it appeared as if the phenomenon of beta decay contradicted the basic conservation laws of physics. The energy spectrum of beta particles is completely different from that of alpha particles: it is not composed of discrete energy values but rather is a continuous curve as shown in Fig. 3.3 (the beta decay of bismuth 210). The energy of a single beta particle may be low, almost nil, or may reach a certain maximal value. Calculations show that the mass difference between the original and produced nuclei is equal to this maximal energy value, yet only a small minority of beta particles actually possess this energy. Most beta particles are emitted with less energy than indicated by the actual mass difference and *are not* accompanied by a photon which would compensate for the energy discrepancy. The beta emission from tritium is one such example:

$${}_1H^3 \rightarrow {}_2He^3 + e^-$$

The mass difference (between ${}_1H^3$ and the sum of ${}_2He^3$ and the electron) is equivalent to 19 500 electronvolts, and this is indeed the maximal energy in the emission spectrum, and yet the *average* energy of the electrons is but half of this value! A similar spectrum was found for all beta emitters, whether they emitted electrons or positrons, and it appeared as if energy was not conserved during beta emission.

An examination of the spins of the emitting nucleus and of its products indicated that angular momentum is not conserved in this process either. Let us consider the most simple beta decay: a free neutron is converted into a proton and an electron. Each of the three particles possesses a spin of $\frac{1}{2}$. The overall spin before the decay is $\frac{1}{2}$, and after the decay is either one (if the spins of the electron and proton are parallel) or zero (if they are antiparallel). In any case, it seems as if angular momentum of $\frac{1}{2}h/2\pi$ either disappears or is created! In 1931, even before the neutron was discovered, the Austrian physicist Wolfgang Pauli (who formulated Pauli's principle, mentioned above) proposed a simple idea in order to explain the energy spectrum of beta particles: it seems, Pauli contended, that during the process of beta decay an additional particle is emitted. This particle carries part of the energy liberated in the process, yet it cannot be detected by ordinary apparatus. This mysterious particle would have to be electrically neutral, because the beta particle takes all the charge missing in the nucleus. Its mass would have to be very small, or even zero, since the most energetic beta particles possess all the energy lost by the nucleus, or almost all of it.

Pauli's proposal was a starting point for a new theory. When the neutron was discovered in 1932 and it was unequivocally proven that there are no electrons in the nucleus, it became evident that any theory of beta decay would have to account for the missing spin as well, and to provide some kind of explanation for the process by which a neutron is converted into a proton and an electron (or a proton into a neutron and a positron). The picture was accomplished by Fermi in 1934. He constructed a theory based on the following assumptions: in addition to the force which binds the nucleons to one another in the nucleus, there exists another force which is capable of converting a nucleon from the neutron state into the proton state, and vice versa, accompanied by the emission of an electron or positron. (This force was eventually named the 'weak force' as opposed to the 'strong force' which binds the nucleons to each other.) Along with the electron, an additional particle is created of spin $\frac{1}{2}$, with zero charge and *zero rest mass* and this particle carries the 'missing' energy and momentum. Fermi named this particle the 'neutrino' (meaning 'tiny neutron' in Italian). Fermi also constructed a physical model based on these assumptions and demonstrated that it provided clear explanations for the energy distribution of beta emitters, as well as for the relationship between the halflife of the nucleus and the energy liberated during the emission – provided that two particles are emitted in the process and not one!

It is interesting to note that the journal *Nature* returned the paper that Fermi had sent, in which he presented his theory, claiming that the paper was 'too speculative' (this story has consoled many a scientist whose paper was not accepted for publication). Fermi, disappointed, decided that he had had enough of theoretical work, and turned to experimentation. During the same year he discovered how to induce artificial radioactivity by neutron bombardment. His paper on the neutrino finally found its way into a German journal.

The success of Fermi's theory provided circumstantial proof for the existence of the neutrino but all attempts to discover it experimentally failed, and the hypothetical particle was nicknamed 'the little one who was not there'. It was only with the construction of nuclear reactors which emit an abundance of neutrinos were physicists finally able to obtain direct experimental evidence for the existence of this elusive particle. Two physicists, Clyde Cowan and Fred Reines, from the laboratories of Los Alamos in the United States provided the final proof. The two began to plan their experiment in 1953, and chose a nuclear reactor in South Carolina as a source of neutrinos.

Inside a nuclear reactor, the uranium decays steadily into beta-emitting isotopes. If beta decay is indeed accompanied by neutrino emission, reasoned Cowan and Reines, neutrinos must be continuously emitted by the reactor in vast quantities. And when this gigantic flux of neutrinos passes through a detector, specially designed for this purpose, some of them may be trapped, identified, and their actual existence verified.

One of the difficulties facing the architects of this experiment was distinguishing between the neutrinos and background radiation emitted from the reactor, i.e. the flux of all other particles that were not neutrinos. It was thus necessary to find a reaction which could be initiated only by the neutrino, and to use this as the sole basis for its identification.

The reaction selected was the reverse process to beta emission. In the process of beta emission, the neutron breaks down into a proton, an electron and a neutrino. Various considerations showed that the particle emitted in this process is most conveniently defined as the anti-neutrino. The process is thus written as:

$$\mathrm{n} \rightarrow \mathrm{p} + \mathrm{e}^- + \bar{\nu}_\mathrm{e}$$

The Greek letter ν (nu) represents the neutrino; the bar over it – the fact that it is an anti-neutrino. The letter e to the right of the ν denotes the fact that it is an 'electron-type' neutrino (as we shall see later on, there are other types of neutrino).

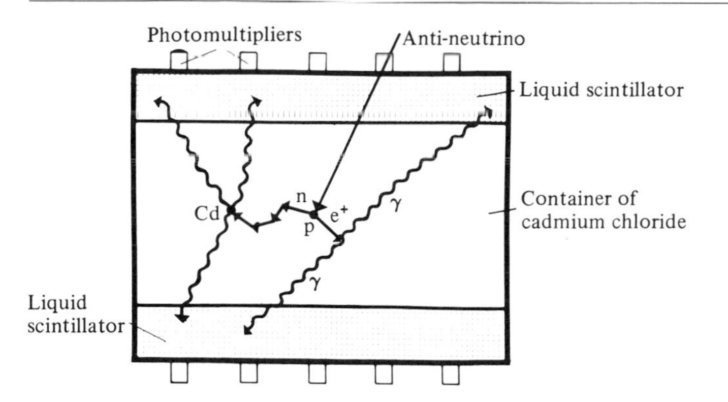

Figure 3.4. A schematic description of the processes in the neutrino detector. When an anti-neutrino is absorbed by a proton, the proton emits a positron (e^+) and turns into a neutron (n). In the annihilation of the positron with an electron, two photons are produced. The neutron is absorbed by a cadmium nucleus which subsequently emits several photons. The photons are detected by liquid scintillators surrounding the container.

The reverse process takes place (with very low probability) when an anti-neutrino collides with a proton:

$$\bar{\nu}_e + p \rightarrow n + e^+$$

The proton emits a positron and turns into a neutron.

The detector which was developed by the 'neutrino hunters' was based on a container full of cadmium chloride solution. Fig. 3.4 illustrates the course of events when an anti-neutrino which has penetrated the container is absorbed by a proton (a hydrogen nucleus in the water in which the cadmium chloride is dissolved). The emitted positron undergoes annihilation with an adjacent electron, and two gamma photons, each possessing a characteristic energy of 0.51 MeV, are produced. The neutron generated in the interaction has initial momentum, but after a number of collisions with water molecules it slows down and is trapped by a cadmium nucleus. As a result, the cadmium nucleus emits a number of gamma photons with a total energy of 9 MeV. Liquid scintillators were installed above and below the container: these are large containers, containing a liquid whose atoms are easily ionized by gamma rays and thus emit light. More than a hundred photomultipliers were installed around the scintillators, to detect and record the scintillations. The number of scintillations is an indication of the total energy of the photons passing through the scintillators. Thus, if the photomultipliers detect two simultaneous gamma photons, each with an energy of 0.51 MeV, and a few thousandths of a second later a number of photons with a total energy of 9 MeV, it is clear beyond all doubt that an anti-neutrino has penetrated the container, like a thief who breaks into a house but leaves behind fingerprints which are unmistakably his.

The planners of the experiment had to surround the entire set-up with a thick layer of earth and metal in order to shield it from penetration by other particles emitted from the reactor and by cosmic rays. (Veteran World War II battleships, sold off for scrap, saw further action in experiments on neutrino research.) As for the neutrino, it passes through all the protective layers as if they were not there at all, because its interactions with matter are so rare. In order to stop a beam of neutrino particles completely, we would need a wall more than 3×10^{16} kilometres, or about 3500 light years, thick! This thickness is 200 million times the distance between the sun and earth. However, *some* of the particles in a neutrino beam may be stopped within a smaller distance than that; the hopes of the 'neutrino hunters' were based on this tiny fraction of the particles.

After 3 years of work their efforts were crowned with success, and in 1956 Cowan and Reines reported the discovery of the neutrino. The tremendous flux of 10^{13} neutrinos per square centimetre per second emitted by the reactor yielded only three reactions per hour that could be detected by the apparatus. However, when the reactor was not in operation, the number of detected reactions dropped to zero. This proved that the neutrinos originated within the reactor and not, for example, in cosmic rays. During the 1960s, large accelerators were constructed (see Chapter 4) which provided powerful sources of neutrino particles, and aided the investigation of the properties of this particle. The mass of the neutrino, according to all indications, is very small, perhaps zero. It possesses a spin of $\frac{1}{2}$, and reactions between it and matter are exceedingly rare. Such reactions are mediated solely by the weak force. The anti-neutrino accompanies any electron emission from the nucleus. For example, the beta decay of tritium should be written:

$$_1H^3 \rightarrow {_2He^3} + e^- + \bar{\nu}_e$$

Some radioactive isotopes decay by emitting positrons and neutrinos, for example:

$$_7N^{13} \rightarrow {_6C^{13}} + e^+ + \nu_e$$

During the 1930s a particle named the muon (described later) was discovered. It was found that in some of the processes in which the muon participates, a neutrino, similar in its properties to the neutrino of the electron, is also produced. However, in 1962 the

Americans Lederman, Schwartz and Steinberger found that the muon neutrino is in fact different from the 'electron' neutrino – it 'remembers' its origin and in reactions which it initiates only muons, and not electrons, are produced. The experiment is described in section 4.3.

Today the neutrino is no longer so hard to find. In modern accelerators we can produce rather dense beams of neutrinos and use them to investigate interactions between the neutrino and other particles. These studies contributed to our understanding of the weak force and even shed light on the internal structure of such particles as the neutron and proton, as we shall see later on.

A new neutrino detector based on gallium has been recently installed in the Gran Sasso tunnel in central Italy. This detector is sensitive to the low energy neutrinos emitted in the basic fusion reaction in which hydrogen is turned into helium. The previous detectors could only sense the more energetic neutrinos accompanying the subsequent nuclear reactions in which heavier elements are produced.

An exciting chapter in the story of the neutrino is the 'neutrino telescope' for solar study. Each 'telescope' of this kind is a neutrino detector similar in principle to the one we previously described. In order to shield the detector from background radiation, it is usually located in an abandoned mine, hundreds of metres below the earth's surface. The layers of rock and earth absorb the cosmic rays but are not able to stop the neutrinos arriving on earth from outer space. Most of these neutrinos originate in nuclear reactions within the sun. Measurements which have been made using the 'neutrino telescope' have led several scientists to the hypothesis that there is a periodical decrease in the activity of the natural 'nuclear reactor' within the sun which may bring about a new ice age in the coming century.

Experiments have been carried out in the United States in an attempt to exploit neutrino beams as a method of communication with submarines. The idea is simple: one can aim a beam of neutrinos which has been produced in an accelerator at a submarine even if it is on the other side of the globe. The neutrinos pass right through the earth as if it were empty space. A portion of the neutrinos will interact with the protons in the water and cause characteristic scintillations in the vicinity of the vessel. By modulating the beam, it is possible to transmit messages, for example by Morse code. Since this method has only military implications, the results of these investigations are being kept secret.

3.4 Yukawa's theory of the strong force

In 1935 the Japanese physicist H. Yukawa published his theory on the nuclear force which binds the nucleons in the nucleus to one another (later to be known as the strong force). This theory included a simple mathematical expression for the dependence of this force on the distance separating two nucleons. It was then known that the nuclear force is short-ranged – its influence is felt only between two nucleons whose centres are no more than 2 or 3 fermi apart (1 fermi $= 10^{-13}$ centimetre). Yukawa proposed a formula identical to that of the electric force, but multiplied by an additional factor which becomes extremely small when r (the distance between the centres of the two nucleons) exceeds a certain limit. Up to this point it was just a simple exercise in mathematics, but here Yukawa introduced a daring and sophisticated step. It is known that when an electrically charged particle accelerates or decelerates it emits photons. Yukawa claimed that just as the electric field can produce photons, the field of the strong force which prevails around the nucleon must also produce, under certain conditions, some kind of particles. Yukawa proceeded to predict the mass which such particles might possess. The only experimental fact at his disposal was the range of nuclear forces (i.e. the distance at which the force became negligible). But even with this fragment of information Yukawa was able to predict (with considerable success, as we shall see) the mass of these particles which had not yet been detected by any apparatus whatsoever. At first he developed a wave equation for the nuclear force. This equation was similar to the electromagnetic wave equation; it was based, however, on the mathematical expression which he had previously proposed for the dependence of nuclear force on distance. He then applied the laws of Planck and de Broglie, which state that a wave of frequency f and wavelength λ carries energy in quanta of $E = hf$ which have the momentum $p = h/\lambda$. These calculations showed that as opposed to electromagnetic fields whose quanta are of zero rest mass, in the case of the strong interaction the mass of the quantum differs from zero, and is equal to:

$$m \text{ (MeV)} = \frac{hc}{2\pi r_0}$$

r_0 represents the range of the force. If we now substitute

$$r_0 \approx 2 \times 10^{-15} \text{ metres}, \quad c = 3 \times 10^8 \text{ metres per second}$$

$$\frac{h}{2\pi} \approx 10^{-34} \text{ joules} \cdot \text{second} = 6.55 \times 10^{-22} \text{ MeV} \cdot \text{second}$$

we obtain: $m \approx 100$ MeV, which is about 200 times the mass of the electron or one-tenth of the proton mass. We can arrive at precisely the same equation by using relatively simple mathematical considerations, which are based on a re-evaluation of the concept *force*.

What is force?

Two electrical charges at rest exert an electrostatic force on one another, just as any two masses exert a gravitational force on each

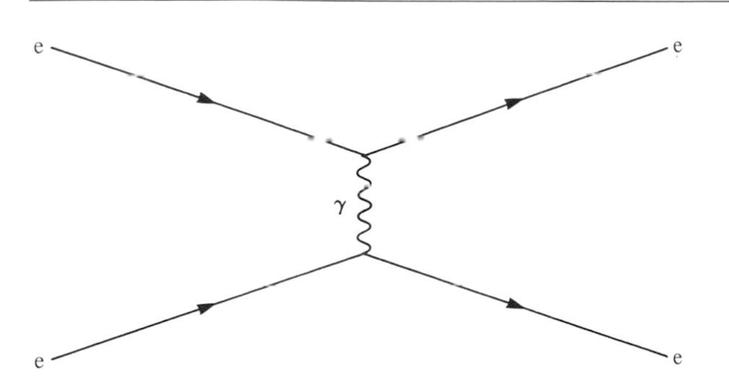

Figure 3.5. According to modern physics, the electric repulsion between two electrons is caused by the exchange of photons. The process is schematically depicted by this 'Feynman diagram'.

other. What transfers these forces from one body to another? In other words, how does one body 'know' of the existence of the second body, removed from it in space? We say that the electric charge is surrounded by an 'electric field' or 'electromagnetic field' which influences the second charge (similarly, mass gives rise to a 'gravitational field'), but is this explanation sufficient? If we look for the definition of an electric field in books on electromagnetism, we will find that the field at a certain point is defined as the force acting on a unit of charge at that point. This is a simple mathematical definition which doesn't answer the question: what transfers this force from one object to the other?

In the twentieth century this question has found an astounding answer. Modern physics tells us that the effect of the electric force is transmitted between charged bodies by the mutual exchange of photons. According to Maxwell's equations in the nineteenth century, photons are emitted by an accelerated electric charge. The modern viewpoint, on the other hand, holds that even an electric charge at rest (or an electric charge moving with a fixed velocity) emits photons. A photon emitted by a charged particle might be absorbed by another charged particle and in that way momentum is transmitted from one particle to another, which is the same as saying that the two particles exert a force on each other.

It is easy to understand how the exchange of photons generates a repulsive force: think of two persons who play with a heavy ball by passing it to each other. Every time one of them catches the ball, he feels that he is pushed backward. It is more difficult to explain by this model the attractive force existing between opposite charges. (The nearest perceptible picture comparable to that situation is perhaps two atoms in a molecule, attracted to each other due to shared electrons which are alternately in one atom and the other.)

However, within the framework of quantum electrodynamics, the attractive electric force can be explained as well.

A more troubling question is the following: why can't we detect – even with the most sophisticated apparatus – photons emitted from a charge at rest? And how can a charged body emit photons without losing either mass or energy? The answer lies in the uncertainty principle. In order for us to be able to detect a photon of energy E, it must exist for at least the time Δt where $\Delta t \times E \approx h/2\pi$. Calculations show that a photon of green light (possessing an energy of 2 MeV) can be detected only if it exists for longer than 10^{-15} seconds. A resting charge does not emit ordinary photons, but 'virtual' photons. They exist for only as long as the uncertainty principle permits them to exist without being detected. After which they 'melt' away or are absorbed by their source unless they previously encountered another charge and were absorbed by it.

Because of the uncertainty principle we are also not able to detect any loss or fluctuations in the mass of a resting charged body due to emission of virtual photons, and of course we cannot detect the virtual photons themselves. The less energetic the virtual photon, the longer it can exist and the farther it can travel. This is the reason that the electric force between charges decreases with distance.

This new outlook views resting particles such as the electron not just as packets of static charge, but as a sort of octopus, continuously sending out long arms at an unimaginable rate, feeling out towards neighbouring charges. The effect of gravitational force, according to this theory, is also transmitted by particles called gravitons; however, these particles have not been discovered experimentally to date, and there is reason to believe that it will be very difficult to do so (see Chapter 7).

The range of electromagnetic force is infinite: its intensity gradually decreases with increasing distance, but at no point does it fall off abruptly. This indicates that infinitely weak virtual photons may be sent by every charge to infinite distances. The strong force, on the other hand, has a finite range since experiments show that beyond a certain distance it drops sharply to zero. One can infer from this fact that the particles which transfer the strong force have rest masses which differ from zero! Since they possess finite mass, they cannot carry an infinitely small energy over an infinite distance. Their minimal energy is actually their rest mass, and thus the lifetime which the uncertainty principle permits them to have is limited. Let us denote this time span Δt. Since the velocity of these particles cannot exceed the speed of light, the maximal distance that they can cover during the time Δt is: $r_0 = \Delta t \times c$, and r_0 is thus the range of the strong force. We will now use the uncertainty principle: $\Delta E \times \Delta t \approx h/2\pi$ where ΔE represents the minimal energy of the particle, i.e. its mass as expressed in units of energy. If we substitute $\Delta t = r_0/c$ and $\Delta E = m$ (MeV), we will obtain m (MeV) $= hc/2\pi r_0$ and this is exactly Yukawa's equation.

3.5 The meson riddle

Yukawa concluded his paper with a doubtful remark, stating that since a particle of the predicted mass had never been detected, his theory was perhaps mistaken. Fourteen years of experimental and theoretical efforts were needed to dispel Yukawa's doubts and show that he was indeed on the right track. In 1949, after the predicted particle was discovered and investigated, Yukawa was awarded the Nobel prize for physics.

New faces in cosmic rays

During the years 1933–36 a number of experimental physicists reported strange findings in cosmic rays. Among the diligent researchers in this field were C. Anderson (the discoverer of the positron) and S. H. Neddermeyer from the USA. They found in cosmic rays positive and negative particles with unusual powers of penetration, which surpassed those of all particles known at the time. At first physicists tended to believe that the particles were electrons and positrons. However, in 1937 it became apparent that in the secondary cosmic rays there are particles heavier than the electron but lighter than the proton. A more precise estimate of the mass of these particles was not at all easy, yet several physicists succeeded in doing so in 1938–39, using cloud chamber photographs. From the curvature of the path of a particle in the cloud chamber, under the influence of a magnetic field, the ratio e/m was calculated. The range of its penetration through matter provided an estimate of its energy, and the thickness of the track of drops in its wake in the chamber (which is a function of its ionizing ability) was an indication of its speed. Various experiments such as those of Neddermeyer and Anderson from the USA and of Blackett (Nobel laureate in 1948) and Wilson from Britain in 1939 gave an average mass of about 100 MeV, which corresponded nicely with Yukawa's predictions. There was no doubt in the minds of the researchers that these particles are the carriers of the strong force discussed by Yukawa, and that they are created when protons of the primary cosmic radiation encounter atomic nuclei in the upper reaches of the atmosphere, just as an electron emits photons when it collides with a nucleus. The new particles were first named mesotrons and later, mesons (from the Greek word for middle or intermediate, since their mass lies between that of the electron and that of the proton). As we shall see later on, this was not yet to be their final name.

Discovering the muon and pion

Measurements of the flux of particles at various altitudes, using Geiger counters, showed that their average lifetime is about two-millionths of a second. According to a naive calculation, such a particle, even if moving at almost the speed of light, should decay after less than one kilometre ($2 \times 10^{-6} \times 300\,000 = 0.6$ kilometres). These so-called mesons, however, could also be detected at sea level. How were they able to traverse the entire atmosphere? The answer lies, of course, in the theory of relativity. We have mentioned (see section 2.1) that at speeds close to the speed of light, time passes much slower, and two-millionths of a second in the coordinate system of the particle are much longer in the eyes of a stationary onlooker. It was found that at the end of its life the particle decays into an electron (or positron) and other neutral particles which do not leave tracks in the cloud chamber; these particles were later identified as a neutrino pair.

The success of Yukawa's theory stirred a lot of excitement, but an intensive look at the facts should have raised a few doubts. These new particles were supposed to be the glue holding all the nucleons together in the nucleus. If so, they should be absorbed, shortly after their formation, by nuclei of atoms in the atmosphere. And yet, their flux at sea level and even at lake bottoms was not appreciably different from their flux on mountain tops. The second world war diverted the interest of physicists to other problems and ten years passed from the time of the meson's first identification (1937) until the discovery that this identification was a mistake from the very start.

The experiments of Conversi, Pancini and Piccioni began during the war, in a cellar in Rome, where the three were hiding in order not to be sent to forced labour in Germany. Despite the wretched conditions in which they worked, they managed to carry out very accurate measurements of the decay of the so-called 'mesons' in various materials. Large budgets and a well-equipped

In 1947, three young Italian physicists, Marcello Conversi, Ettore Pancini and Oreste Piccioni, published results of experiments which were intended to measure the absorption of the positive and negative particles named mesons in various nuclei. The positive particle is not attracted to the nucleus, and thus decays spontaneously. It was found, however, that the negative particle slows down after having collided with several nuclei, and begins to orbit around a specific nucleus and, after one-millionth of a second, is absorbed by the nucleus. This time interval, which seems exceedingly brief to us, is a long time in the particle world, too long in fact. According to Yukawa's theory, the quantum of the strong force should be absorbed by a nucleon within 10^{-23} seconds, and not within 10^{-6}

laboratory are not always necessary conditions for good scientific work (the discovery of radium by the Curies is yet another example).

seconds as observed! This discrepancy of the order of 10^{17} between theory and experiment dramatically confirmed that something was wrong.

A number of explanations were offered in an attempt to save the theory. The explanation which was proven to be correct was the simplest one: *the so-called meson is not Yukawa's particle*. Actually there are two different particles. The particle predicted by Yukawa, from now on known as the pi (π) meson, is indeed created in the upper reaches of the atmosphere by collisions of cosmic particles with nuclei of atoms in the air. The pi mesons do not get very far because the probability of interactions between them and nuclei in the atmosphere is high. The product of these interactions is a particle which was renamed the mu (μ) meson. It is not a carrier of the strong force, and since it does not easily react with matter, it has time enough to penetrate the atmosphere down to sea level. This is the particle which was mistakenly identified for ten years as the particle that carries the strong force. This explanation was provided by Soicho Sakata and Tokuzo Inoue of Japan and Robert E. Marshak and Hans Bethe (1967 Nobel laureate) of the United States, and experimental proof was not long in coming. In due course the name mesons was assigned to particles which transfer the strong force and thereupon the name mu meson was replaced by the name muon, denoted by the Greek letter μ. The pi meson is frequently called the pion, and it is represented by the letter π.

In order to discover the pion experimentally, it was necessary to send detectors to the upper reaches of the atmosphere. This can easily be accomplished by unmanned balloons, for example. The problem was to design appropriate detectors which would function efficiently without human supervision. C. F. Powell and his group at Bristol solved this problem by using sensitive photographic plates and an extraordinary technique of analysis, which included photographing tiny portions of the particle track under the microscope. The mass of the particle and other data can be obtained by measuring the density of the small black dots which form the track, and from deviations of the track from a straight line (due to collisions with nuclei in the emulsion). The slower the particle, the more atoms it is able to ionize, and the greater the density of the dots. Thus, the direction of the particle's motion may be determined (the thicker portion in the track represents the end of the particle's travel) and its energy as well as the *rate of energy loss* estimated. The average deviation from a straight line helps to approximate the mass of the particle.

Important contributions to the use of photographic plates were made by G. Occhialini, who in 1945 joined Powell in Bristol. In 1979 Occhialini shared the Wolf prize in physics (with G. Uhlenbeck) for his part in the discovery of pair production and his role in the discovery of the charged pion.

The photographs which returned from the heavens clearly

Figure 3.6. The paths of a pion (π) of the secondary cosmic rays, the muon (μ) which was produced in its decay, and the electron (e^-) produced in the decay of the muon, as recorded on a photographic plate. The momentum of the neutrinos formed in each decay can be evaluated from the angles between the tracks of the charged particles. (From C. F. Powell, P. H. Fowler & D. H. Perkins, *The Study of Elementary Particles by the Photographic Method*: Pergamon Press, 1959.)

showed the tracks of two kinds of particles. The heavier particle was identified as the pion. At the end of its track another track, that of a lighter particle, began. This particle was identified as the muon. In an even more sensitive emulsion it could be seen that the muon decayed into a light particle which was identified as an electron (see Fig. 3.6). The angle at which the electron was emitted indicated the emission of other, neutral particles, as well (neutrinos). An analysis of these tracks fixed the mass of the muon at 106 MeV and the mass of the charged pion at 140 MeV – a little higher than Yukawa's prediction.

Other photographic evidence for the decay of the pion into the muon and the muon into the electron was provided one year later (in 1948) by the group of C. M. G. Lates of the University of California in Berkeley, but the source of the particles was artificial this time: a particle accelerator which began its operation at Berkeley the same year and was able to provide protons with kinetic energies of up to 200 MeV.

In the next chapter you will read of the development of accelerators and their effect on particle research.

The neutral pion

Without knowing that the 'mesons' discovered experimentally were not the same particles that Yukawa had predicted, theoretical physicists continued, in the years between 1935 and the second world war, to develop the meson theory. In 1938, the British physicist Nicholas Kemmer published his theory of charge symmetry. According to this theory, the fact that the strong forces between proton and proton, proton and neutron, and neutron and neutron are similar shows us that there must be positive, negative and neutral mesons.

The neutral pion was more difficult to discover than its charged brothers, because of its much shorter life span and because neutral particles do not leave tracks in the detectors. However, in 1950 the neutral pion was positively identified by detecting and analysing its decay products.

Thus, after 15 years of research, Yukawa's theory finally found comfort and grace. Meanwhile, the particle zoo had been enriched by a number of new members. The meson story did not pass, however, without leaving behind its wounds and bruises. The Swiss physicist, Baron Ernest C. G. Stueckelberg developed a theory identical to that of Yukawa in 1935. He showed his work to Wolfgang Pauli who ridiculed it. Pauli generally tended to be critical of new ideas. Stueckelberg was dissuaded from publishing his ideas, and thus did not share the Nobel prize which was awarded to Yukawa in 1949.

Another story of personal anguish is connected with the meson episode. The Japanese scientists Sakata and Inoue had already published their 'theory of two mesons' in 1943 in Japan. The English translation was submitted for publication only in 1946, and soon after, a similar theory was presented, independently, by the Americans Marshak and Bethe. Because of the war, the work of the Japanese attracted the attention of Western scientists only about a year later. Thus the mistaken impression grew that the Americans had got there first, and in articles published later no mention was given of the contribution of the Japanese, but only of the American pair. This, in turn, caused great bitterness in Japan. The bitterness exploded at an international physics conference held in Kyoto, Japan, in 1965, on the 30th anniversary of Yukawa's theory. M. Taketani, one of Japan's most famous physicists, who bore a personal grudge against the United States and her scientists, accused the latter of plagiarism. Marshak was present at the conference and in his touching answer he publicly gave the credit to the Japanese. The verbal exchanges were all published in the conference proceedings.

3.6 The properties of the pions

The pion and the muon with which we made acquaintance in the previous section are completely different from one another in their properties. Their masses are similar, however, and thus a few years of confusion passed until they were distinguished from one another.

Today we know that they are members of two different groups of particles, just as the bat and raven belong to two different classes of the animal kingdom, despite the similarity between the two.

The muon is similar to the electron in being completely unaffected by the strong force. We shall see later that both belong to the same 'family'* (the lepton family). The pion on the other hand, belongs to another family, the meson family, and was in fact the first member of this family to be studied by physicists.

It is quite easy to produce pions and muons in particle accelerators, as we shall see in the next chapter. Their lifetimes, which seem very short in everyday terms, are actually quite long in terms of the particles world. Once produced in an accelerator it is easy to investigate them and their properties are now known to a great precision. In this section the pions' properties are surveyed, and in the next one – those of the muons.

The pions make up a triplet: a positive (π^+), a negative (π^-) and a neutral (π^0) particle. Their important properties are summarized in Table 3.2. The two charged pions are a particle–anti-particle pair: both have exactly the same mass and average lifetime (when allowed to decay spontaneously). The neutral π^0 on the other hand is its own anti-particle. Note that its mass is slightly less than that of its charged brothers, and its lifetime is shorter by eight orders of magnitude than the lifetime of a charged pion. The small difference between the masses and the large difference in the lifetimes stem from the neutral pion's lack of charge. All three pions have zero spin.

Pions are formed in great abundance as the result of collisions between nucleons. The collision between a rapidly moving proton (in cosmic rays or in an accelerator) and a proton at rest, may give, for example, the following results:

$$\mathrm{p} + \mathrm{p} \rightarrow \begin{cases} \mathrm{p} + \mathrm{p} + \pi^0, \text{ or} \\ \mathrm{p} + \mathrm{n} + \pi^+, \text{ or} \\ \mathrm{p} + \mathrm{p} + \pi^+ + \pi^-, \text{ etc.} \end{cases}$$

The number of pions formed increases with the energy of the incident proton. The formation of two pions is a characteristic process at energies in the vicinity of 2000 MeV, while at very high energies more than 100 pions can be produced in a single collision. (The tracks in Fig. 3.1 are those of pions.)

* We use the term 'family' for particles with similar properties. Note, however, that since 1978 the same term has sometimes been used in a totally different context, for which we use in this book the more appropriate term 'generation' (see section 10.9).

Table 3.2. *The pions*

Particle	Electric charge (in *e* units)	Spin	Mass (MeV)	Average lifetime (seconds)	Main decay mode
π^+	+1	0	139.6	2.6×10^{-8}	$\pi^+ \rightarrow \mu^+ + \nu_\mu$
π^-	−1	0	139.6	2.6×10^{-8}	$\pi^- \rightarrow \mu^- + \bar{\nu}_\mu$
π^0	0	0	135.0	0.9×10^{-16}	$\pi^0 \rightarrow \gamma + \gamma$

Collisions between a proton and a neutron may also produce pions, for instance:

$$p + n \rightarrow p + p + \pi^-$$

It is easy to understand these processes with the help of the model which describes the strong force as carried by 'virtual' mesons. According to this model a proton or neutron at rest continuously emits virtual pions which are reabsorbed after travelling an average distance of 1 fermi (10^{-13} centimetres). These virtual pions cannot be converted into 'real' pions and leave the nucleon, due to the law of energy conservation. However, when a rapidly moving nucleon passes close to another nucleon, it passes through its 'cloud' of virtual pions and may tear some of them away, that is to say, provide them with sufficient energy to turn into real particles! The greater the energy of the incident nucleon, the greater the number of pions it can liberate. This energy is, of course, at the expense of the kinetic energy of the nucleon, which subsequently slows down. A schematic description of the process appears in Fig. 3.7.

Photons, electrons and rapidly moving pions can also produce pions when colliding with a nucleon at rest. The following processes may, for example, take place:

$$\gamma + p \rightarrow \begin{cases} p + \pi^0 \\ p + \pi^+ + \pi^- \\ n + \pi^+ \end{cases}$$

When a pion collides with a nucleon, processes of 'charge exchange' may occur in which the nucleon is transformed from a proton to a neutron or vice versa.

$$\pi^- + p \rightarrow n + \pi^0$$
$$\pi^+ + n \rightarrow p + \pi^0$$

The main decay process of a charged pion (π^- or π^+) results in the formation of a muon, accompanied by a muonic neutrino. (Most of the muons in cosmic rays originate in this manner.) Occasionally the

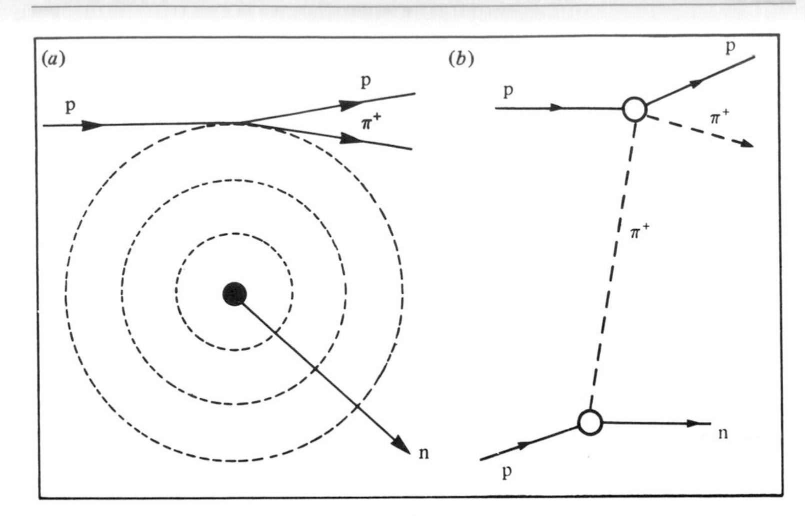

Figure 3.7. (a) The reaction $p + p \rightarrow p + n + \pi^+$ can be regarded as the tearing of a pion from the cloud of virtual pions wrapping the proton. (b) The Feynman diagram of that process.

decay products also include a photon. Once in about 8000 cases, the decay results in electron or positron formation (and an electronic neutrino).

$$\pi^+ \rightarrow e^+ + \nu_e \quad \pi^- \rightarrow e^- + \bar{\nu}_e$$

The decay processes of the charged pions resemble beta decay. As we shall see later, the two processes have much in common. The π^0 usually decays into two photons: $\pi^0 \rightarrow 2\gamma$. (This decay mode is of course not possible for the charged pions due to charge conservation.) These photons are capable of producing electron–positron pairs, and sometimes one or both of them do so even before being separated from the decaying pion. The process we then observe is:

$$\pi^0 \rightarrow \gamma + e^+ + e^- \text{ or } \pi^0 \rightarrow 2e^+ + 2e^-$$

The first process occurs in 1 out of 100 π^0 decays approximately, and the second – in 1 out of 30 000.

In Chapter 6 concerning the four basic forces, we will be able to understand why the decay of the π^0 is much faster than the decay of the charged pions.

3.7 The properties of the muons

The muon is very similar to the electron, and may be viewed in all aspects as an 'overgrown' electron. Its mass is 105.6 MeV (207 times greater than the electron mass). It has a spin of $\frac{1}{2}$, carries positive or negative charge (but not zero!) just like the electron, and has almost

Table 3.3. *The muons*

Particle	Electric charge (in e units)	Spin	Mass (MeV)	Average lifetime (seconds)	Main decay mode
μ^+	+1	$\frac{1}{2}$	105.6	2.2×10^{-6}	$\mu^+ \rightarrow e^+ + \nu_e + \bar{\nu}_\mu$
μ^-	−1	$\frac{1}{2}$	105.6	2.2×10^{-6}	$\mu^- \rightarrow e^- + \bar{\nu}_e + \nu_\mu$

the same ratio of spin to magnetic moment. Moreover, when formed in a pion decay the muon is accompanied by a neutrino, exactly as is the electron when formed in a neutron decay (beta decay). μ^- and μ^+ are particle and anti-particle (just like e^- and e^+). Their important properties are summarized in Table 3.3 (ν_μ and $\bar{\nu}_\mu$ are the muonic neutrino and anti-neutrino).

The muons are not affected by the strong force, and as they pass through the air following their formation high in the atmosphere, they lose energy only by electromagnetic interaction with the nucleus and electrons in the atom. The main process which slows them down is the energy loss due to the ionization of atoms which they encounter. Despite this slowing down, a considerable number of muons reach sea level, and the flux of muons at that altitude is, on average, one per square centimetre per minute.

When the negative muon is stopped within matter before it has the opportunity to decay spontaneously, it is attracted to a nucleus and for about a millionth of a second it goes in an orbit around the nucleus, just as if it were an electron. During this period it cascades down through orbits closer and closer to the nucleus, and at each jump a photon is emitted. Studies of these photons provided important information both on the muon and on the nucleus itself. The muon may remain in the innermost orbit until it either decays spontaneously or is gobbled up by the nucleus (similar to the radioactive process of electron capture):

$$\mu^- + p \rightarrow n + \nu_\mu$$

The positive muon is not attracted to the nucleus and its decay is always spontaneous.

There are other proofs that the electron and muon are identical in all their properties except mass. One of these is associated with magnetic moments. We have already mentioned the fact that the magnetic moment of the electron can be determined by using the equations of quantum electrodynamics and the result (1.001 159 6 μ_e where μ_e is the Bohr magneton) was confirmed by experiment. Similar calculations showed that if the mass of the electron was 207 times as great, its magnetic moment would be 1.001 165 μ_e (but then μ_e would be 207 times smaller). Experiments proved that this is the magnetic moment of the muon, to a precision of seven decimal places! The magnetic moments of other particles, on the other hand, could not be evaluated by the same equations.

The following question might be asked: if the electron and muon are so similar, why is the former a stable particle which never departs from this world unless it is forced to do so (by annihilation with a positron, for instance), while a muon at complete rest decays after 2.2×10^{-6} seconds, on average, to an electron (or a positron) and two neutrinos? This problem has a simple explanation. The electron is stable because it has nothing in which to decay! Being itself the lightest charged particle, it cannot decay into lighter particles (photons or neutrinos) due to the law of conservation of charge. And of course it cannot decay into a heavier particle because of the law of conservation of energy, since where would the extra mass come from?

The great similarity between the muon and the electron was, and still is, a great puzzle to physicists. They could not understand why two particles, so much alike, have such different masses. Moreover, the other particles which we have met so far filled important gaps in the physicist's picture of the world. The discovery of the pions proved Yukawa's theory of the strong force. The appearance of the positron verified Dirac's equation and by detecting the neutrino the apparent loss of energy and spin in beta decay was explained. In this almost perfect picture of the world, no role was found for the muon, hence the amusing protest of the American physicist I. Rabi (1944 Nobel laureate) who declared, when the muon was at last identified correctly: 'Who ordered this?'

As we shall see later (Chapter 10) a third, heavier brother of the electron and muon was found at the end of the 1970s. Its mass is about 3500 times as great as the mass of the electron. We shall see that the attempts to explain why these three similar particles differ so much in their masses are interlaced with the efforts to formulate a comprehensive theory of particles and forces which would explain the properties of all the particles in terms of a small number of basic building blocks.

Chapter Four

Particle accelerators – or from hunters to farmers

4.1 The first accelerators

In the 1930s and 1940s, cosmic rays were the only source of new particles. The high-energy cosmic particles are excellent projectiles for splitting nucleons and producing new elementary particles, and physicists indeed made good use of them for this purpose. However, the disadvantage of this source of radiation is that it cannot be controlled. All that can be done is to send instruments or photographic films into the upper atmosphere and wait for results. If one wishes to investigate a certain process, occurring at a particular energy, one has to sort through a very large number of photographs in order to find a few containing the required information, if at all. As the research of the particle world strode forwards, physicists wanted to imitate in the laboratory the processes produced by the primary cosmic particles. For this purpose they had to accelerate charged particles – protons and electrons – to high speeds with the aid of electric and magnetic fields. This transition 'from hunters to farmers' took place in the 1940s, although its beginnings were in the early 1930s.

The first machine for accelerating particles was built over the years 1928 to 1932 in Rutherford's laboratory (the Cavendish Laboratory) at Cambridge. It was designed and constructed by John D. Cockcroft – an electrical engineer who had come to Cambridge to take a PhD in physics and became one of Rutherford's assistants – and Ernest T. S. Walton from Ireland. Their accelerator consisted of a large transformer and a current rectifier which generated a voltage of several hundreds of thousands of volts between two electrodes. Protons, produced by ionizing hydrogen atoms, were liberated near the positive electrode and arrived at the negative electrode with energies of up to 0.75 MeV (750 000 electronvolts).

When the element lithium was bombarded with these protons its nucleus split into two helium nuclei. For this work Cockcroft and Walton were awarded the Nobel prize in 1951.

In 1930 the young American physicist Robert J. Van de Graaff, who studied in Paris (under Marie Curie) and at Oxford, invented a new method of charging large metal balls to very high voltages, and constructed the particle accelerator named after him.

In the early 1930s another American, Ernest O. Lawrence, invented an accelerator in which particles moved in circles under the influence of a magnetic field, while their velocity was increased in each cycle due to an accelerating electric field. This machine, which was called the 'cyclotron', was a turning point in the advance of particle accelerators, since it made it possible to accelerate particles to much higher energies than those obtained in other accelerators, and at a moderate cost.

The first cyclotron built by Lawrence had a diameter of a few centimetres. The successive ones were larger, and by 1939 he had constructed at Berkeley a 1.5-metre cyclotron weighing 200 tonnes. Similar machines were built in other universities, and they opened a new epoch in particle research. They also caused a breakthrough in nuclear physics, and were soon utilized in the mass production of artificial radioactive isotopes for applications in medicine, agriculture and biological research. Modern particle accelerators are based on similar principles to those of the first prototypes but they are much bigger. The circumference of a modern accelerator might be several kilometres. It may require several years to construct and cost a huge sum of money. The principles of modern accelerators are the subject of the next section.

4.2 The structure of a modern particle accelerator

Particle accelerators are machines which provide charged particles with high kinetic energies by accelerating them in an electric field. The charged particles (protons or electrons usually) enter the accelerator with low velocity and kinetic energy, to be ejected from the other end with high kinetic energy (and a velocity close to that of light, in modern accelerators). The fundamental principle employed in all accelerators is the one used to accelerate electrons in the tube depicted in Fig. 1.1, or in any modern television or X-ray tube. The charged particle is moving between two metal plates (electrodes) which carry electric charges of opposite signs. The electric field prevailing in the region between the electrodes applies a force on the particle. The force is in the direction of the field (i.e. towards the negative electrode) if the particle is positively charged, and in the opposite direction in the case of a negative charge (see

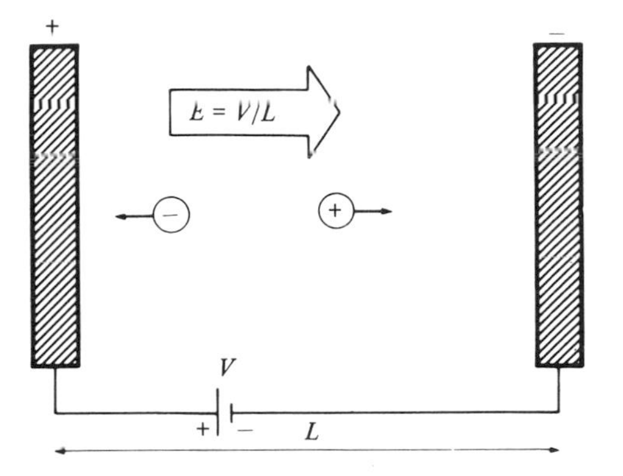

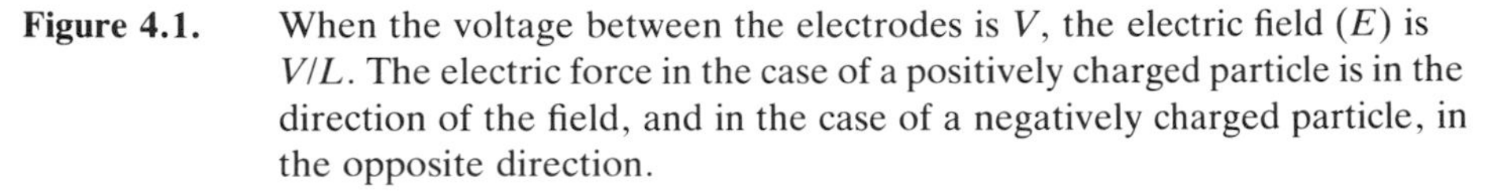
Figure 4.1. When the voltage between the electrodes is V, the electric field (E) is V/L. The electric force in the case of a positively charged particle is in the direction of the field, and in the case of a negatively charged particle, in the opposite direction.

Fig. 4.1). Under the influence of that force the particle is accelerated, that is its velocity is continuously increased.

The magnitude of the electric field is equal to the electric voltage across the electrodes divided by the distance separating them. If, for example, the voltage is 100 000 volts, and the distance between the electrodes is 2 metres, the electric field is 50 000 volts per metre everywhere between the electrodes. To find the electric force acting on the particle, one has to multiply the magnitude of the electric field and the charge of the particle. The energy acquired by the accelerated particle equals the product of the electric force and the distance travelled by the particle under the influence of the force. A simple calculation shows that the total energy gained by a particle which traverses the whole distance between the two electrodes depends only on the *voltage* between the electrodes and on the charge of the particle. It is actually the product of these two quantities:

$$\begin{bmatrix}\text{kinetic}\\ \text{energy}\end{bmatrix} = [\text{invested work}] = [\text{force}] \times [\text{distance}]$$

$$= \begin{bmatrix}\text{charge of}\\ \text{the particle}\end{bmatrix} \times [\text{electric field}] \times \begin{bmatrix}\text{distance between}\\ \text{electrodes}\end{bmatrix}$$

$$= \begin{bmatrix}\text{charge of}\\ \text{the particle}\end{bmatrix} \times \frac{[\text{voltage}]}{[\text{distance}]} \times [\text{distance}]$$

$$= \begin{bmatrix}\text{charge of}\\ \text{the particle}\end{bmatrix} \times [\text{voltage}]$$

The kinetic energy of the particle, as well as its total relativistic energy (kinetic energy + energy embodied in the mass), is conveniently expressed in electronvolts (eV). Recall that 1 eV ($=1.6 \times 10^{-19}$ joules) is the energy acquired by a particle carrying one unit of charge (i.e. the charge of a proton or an electron) when accelerated by a voltage of one volt. If, for example, an electron or a proton is accelerated by a voltage of 10 000 volts, its kinetic energy increases by 10 000 eV. Let us also remind you that 10^6 eV are called MeV, and 10^9 eV are called GeV (= Giga eV).

On the face of it, the most plain and direct way to reach higher energies is by raising the accelerating electric field. Indeed, the builders of the early accelerators (of the Cockcroft–Walton and Van de Graaff types, for instance) focused their efforts on searching for contrivances for increasing the voltage between the electrodes. But the static voltage that can be sustained across a pair of electrodes is limited. Above some maximal value an electric discharge occurs by the onset of arcing between the electrodes or by the breakdown of the insulation holding them. Carefully designed electrodes and supreme insulators can retain a voltage of several million volts. But this is not enough if one wishes to imitate the high-energy processes initiated by cosmic rays. For example, in order to produce a pion by a collision between a moving proton and a stationary one, it is necessary to provide the moving proton with a kinetic energy of more than 200 MeV. A static voltage of 200 million volts is not feasible; nevertheless, charged particles can be accelerated not only to 200 MeV, but to energies ten thousand times higher, by using relatively small voltages. This is accomplished by driving the particle through the accelerating voltage many times, with higher and higher initial kinetic energy each time. Thus, instead of one big 'push', the particle experiences many small 'pushes'. This principle of gradual acceleration is applied in all the modern accelerators. These can be classified into two categories: linear accelerators and circular accelerators or synchrotrons.

Linear accelerators

In a linear accelerator, or *linac*, the particle passes through all the acceleration stages just once. It begins at one end of the accelerator, with a low kinetic energy, and emerges from the other end possessing high kinetic energy. A schematic description of a linear accelerator for electrons is depicted in Fig. 4.2. The electrons are emitted by a heated cathode, and go through an initial acceleration by a static voltage of about a million volts. They then flow through several metal tubes, or cavities, connected to an alternating voltage

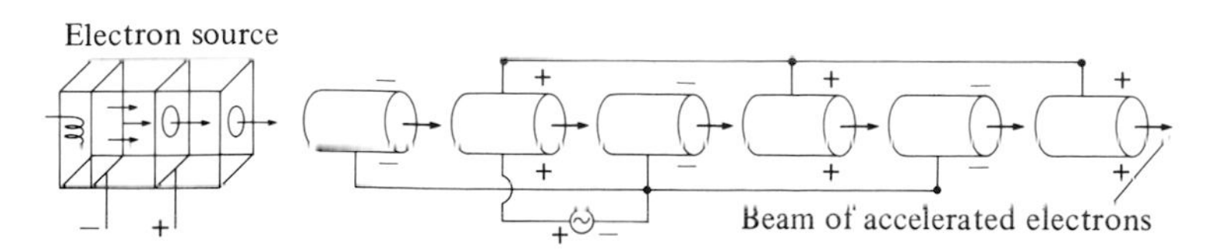

Figure 4.2. Linear accelerator for electrons – a schematic description.

source. The alternating voltage is synchronized so that when the electron beam is inside a tube of positive potential the direction of the voltage is reversed, and the next tube, which was negative, becomes positive.

Thus, when the electrons are between two tubes they are always pushed forward. This system is totally enclosed in a vacuum pipe (not appearing in the figure) so that the accelerated particles will not lose energy by collisions with air molecules.

This arrangement enables the acceleration of electrons to high energies, using moderate electric voltages. If, for example, the voltage between two adjacent tubes is 100 000 volts, and there are 25 tubes in the linac, the overall increment in the kinetic energy of the particle would be 2.5 MeV. In reality, a modern linac may contain hundreds or thousands of accelerating tubes, and its length may reach hundreds of metres or several kilometres. The largest electron linac to date is operating at the Stanford Linear Accelerator Centre (SLAC) near Stanford University in California. It has undergone several modifications throughout the years, but when completed in 1961 it was 3 kilometres long, and contained 82 650 accelerating cavities, which could accelerate electrons to 22 GeV. The energy has later been raised several times and it is now 33 GeV.

The design of a proton linac is similar in principle, but instead of an electron source it has, of course, a source of protons, such as the one described in Fig. 4.3. Another difference between electron and proton linacs is associated with the different mass of the two particles. The mass of the electron is 0.5 MeV, and it takes only a few MeV to bring it close to the speed of light. (When the velocity of a particle approaches the velocity of light, its kinetic energy becomes large compared to its rest mass energy. If we continue to accelerate the particle, the augmentation in energy is mainly associated with the increase of its mass, while the velocity is almost constant.) Thus after passing through the initial static voltage, the velocity of the electron is already close to that of light, and it remains almost constant afterwards. The length and separation of the accelerating tubes are therefore the same throughout the linac. The

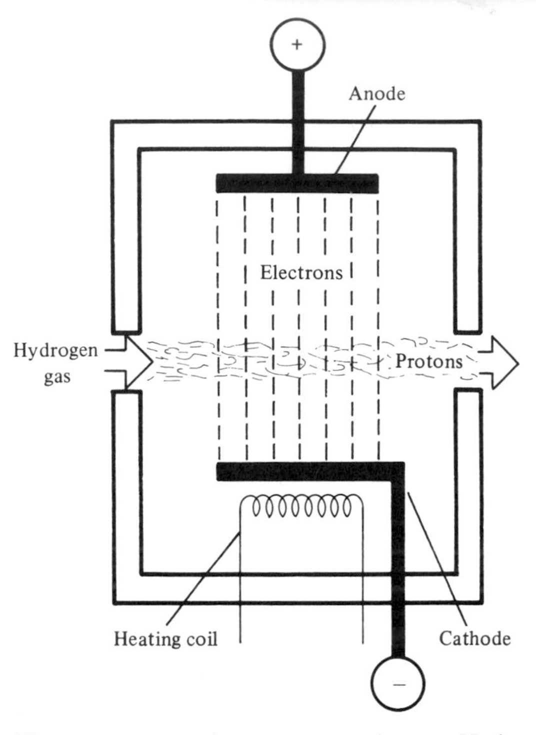

Figure 4.3. The proton source in a proton accelerator. Hydrogen gas is admitted into the cell from the left. As a result of collisions between the gas molecules and fast-moving electrons flowing from the cathode, the molecules break into single atoms which are then ionized and become bare protons.

rest mass of the proton, on the other hand, is nearly 1000 MeV. Its velocity, after the initial acceleration to an energy of several MeV, is still far from the velocity of light, and it is continuously increasing as the proton moves along the accelerator. In order to enable the operation of all the accelerating tubes at the same frequency, despite the change in the velocity of the proton, they have to be shorter at the beginning of the linac and increasingly longer afterwards. Thus, the faster the proton, the longer it has to travel between subsequent spaces, and the frequency of its appearance between the tubes remains constant.

Circular accelerators

In a modern circular accelerator the particles move within a vacuum pipe with the shape of a ring. They complete many revolutions, and gain energy on each revolution. Fig. 4.4 is a schematic description of a circular accelerator for protons (a circular accelerator for electrons is essentially the same). After an initial acceleration in a linac, the

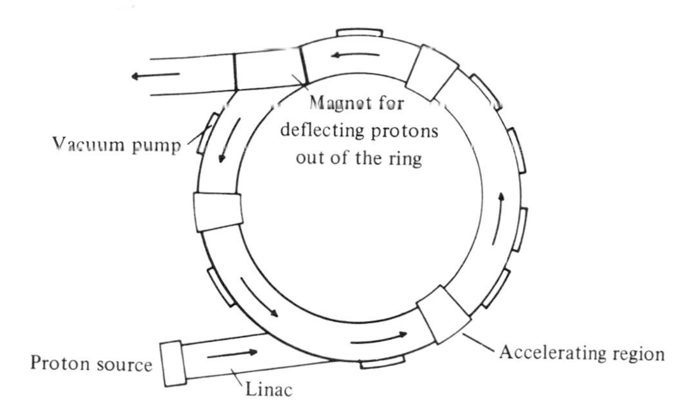

Figure 4.4. A circular proton accelerator. The protons undergo a stage of initial acceleration in a linear accelerator (linac) before entering the main ring. At several locations along the ring there are accelerating electric fields. Focusing and bending magnets along the ring (not appearing in the figure) retain the protons in a narrow beam and constrain them to move in a circular orbit.

proton beam enters the main ring with a kinetic energy of the order of 1 MeV. In one region of the ring (or more) there exists an accelerating electric field, and each time the proton crosses that region, its energy increases. Along the ring, electromagnets of two types are installed. The first type consists of bending magnets which compel the protons to move in a circular orbit. A charged particle moving in a magnetic field is acted upon by a force which is perpendicular both to the direction of the field and to the velocity of the particle. The field of the bending magnet is orthogonal to the plane of the ring, and thus the force on the particle is directed towards the centre of the ring, as shown in Fig. 4.5.

The second type of magnets are called focusing magnets. Their poles are designed in such a way that when the particle beam passes between them, the magnetic field tends to concentrate the beam and restore stray particles to the line. In the early 1950s, magnets which produced 'strong focusing' were contrived; an invention which has had a crucial impact on the development of modern accelerators. Since the particles in the beam exert repulsive forces on each other, the beam tends to broaden and disperse. Thus the first accelerators had to have vacuum pipes with large cross-sections as well as big magnets to contain the pipe between their jaws. Even so, energies higher than 10 GeV were inaccessible since they required a huge number of revolutions in the accelerator, and it was impossible to keep the beam from dispersing throughout the process. With the

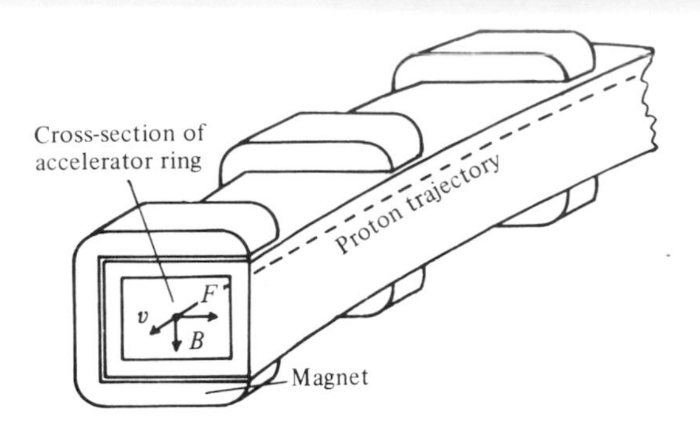

Figure 4.5. The bending magnetic field (B) in a proton synchrotron is directed downwards, and the force (F) is towards the centre of the ring. The velocity of the proton is denoted by v.

strong-focusing method it became feasible to build large accelerators in which the beam remains narrow for hundreds of thousands of revolutions, within a small-bored vacuum pipe with a cross-section of several square centimetres only!

Both the bending and focusing magnets have no influence on the kinetic energy of the protons, which increases only in the region of the accelerating electric field. The higher the energy of the protons, the higher the magnetic field which is required to keep the particles in orbit. The magnetic field is thus increased in each revolution along with the increase in energy of the particles. When the protons reach the desired energy, a special electromagnet is operated which deflects their path and expels them from the ring. At this stage the accelerator is ready to begin the acceleration of another beam.

Because of the required synchronization between the bending magnetic field and the energy of the accelerated particles, an accelerator, in which the radius of the orbit is constant while the magnetic field is gradually altered, is called a *synchrotron*. (In the cyclotron type of accelerator, the magnetic field was constant while the radius of the orbit increased continuously, so that the particles moved in a spiral orbit.)

The maximal energy that can be reached in a synchrotron is limited by two main factors:

The magnitude of the bending magnetic field that can be created.
Energy dissipation due to the electromagnetic radiation emitted by accelerated particles according to the electromagnetic theory.

The last-mentioned phenomenon necessitates a few words of explanation. Motion in a circular path (or any other path which is not a straight line) is an accelerated motion even when it is at a constant speed, since the *direction* of the velocity is continuously changed. Thus, when the charged particles move swiftly in their circular path within the synchrotron, they radiate abundant electromagnetic energy (chiefly in the ultraviolet range of the spectrum) and therefore lose kinetic energy which must be replenished. The emitted radiation is called synchrotron radiation, and it is more severe for the light electrons than for the more massive protons. (The rate of emission is proportional to E^4/M^4R^2 where E is the energy and M the mass of the particle, and R is the radius of the orbit.)

It is interesting that nature itself happens to produce synchrotron-like motions of electrons. The Crab Nebula is a beautiful astronomical object, resulting from a supernova explosion observed by Chinese astronomers in 1054 AD. It emits huge amounts of synchrotron radiation from electrons accelerated by the supernova remnant, a very dense neutron star, rotating some 30 times per second around its axis, and surrounded by a powerful magnetic field, Cosmic rays are apparently accelerated by similar 'accelerators'.

The remedy for both those problems is to increase the diameter of the accelerator ring. The larger the diameter, the smaller the curvature of the orbit and thus a smaller magnetic field is needed to keep the particle on course. In addition, the rate at which the charged particle dissipates its energy by synchrotron radiation is also diminished with the increasing diameter.

The striving towards higher and higher energies has led to the construction of accelerators of ever-increasing dimensions, while the old ones have frequently been converted to facilities for initial acceleration. Each new generation of accelerators has opened up new vistas in particle research, enabling the discovery of new particles and the study of new phenomena. In the 1930s the dimensions of the largest accelerators were a few metres. Those of the 1970s and 1980s, on the other hand, stretch over many square kilometres – usually in underground tunnels – and accelerate electrons and protons to energies of hundreds and thousands of GeV. Their construction and operation cost a fortune and require national or even international efforts.

The first breakthrough into the GeV range took place in the 1950s, with the completion of two large proton synchrotrons in the USA. The first, which was called Cosmotron and reached an energy of 3 GeV, was inaugurated in 1952 at Brookhaven National Laboratory, a new laboratory built on Long Island, New York, through the cooperation of several eastern universities. The second, which was named Bevatron, was activated in 1954 at the University of California at Berkeley, and reached an energy of 6.4 GeV – enough to produce proton–anti-proton pairs. The Russians joined the race in 1955, with a 10-GeV accelerator which had been secretly built at Dubna.

A short time after the inauguration of the Cosmotron and Bevatron, both machines developed difficulties and were stopped for about 6 months. The joke circulated was that the Lord needed that time in order to invent the next puzzles for the physicists and to decide on the correct answers to the previous ones.

The next leap forward occurred in the early 1960s. By 1960 a 28-GeV proton synchrotron had been completed at CERN (Centre Europeén pour la Recherche Nucléaire) near Geneva, Switzerland, by the cooperative efforts of 12 European nations. CERN has since become the most prominent example of an ideal international scientific endeavour. In 1961, a similar accelerator became operative at Brookhaven, with energies of up to 33 GeV, and in 1967 a Russian proton synchrotron with an energy of 76 GeV was constructed at Serpukhov, some 100 km south of Moscow. These accelerators, which dwarfed all their predecessors, look small and modest compared with the accelerators of the 1970s. In 1972 a new large proton synchrotron began operating near Chicago. It had a circumference of 6.3 km and a maximal energy of 400 GeV, which was upgraded to 500 GeV by 1975. The accelerator and the facilities around it has been called the Fermi National Accelerator Laboratory, or Fermilab for short. Before entering the main ring the protons passed through several stages: a Cockcroft–Walton accelerator (which raised their energy to 0.75 MeV), a linac (200 MeV) and a small synchrotron (8 GeV). In the main ring the energy of the protons increased by only 2.8 MeV each revolution, and therefore 140 000 cycles were needed to bring the beam to 400 GeV. However, because of the high velocity of the protons (which was close to the speed of light) the whole process took only a few seconds.

In 1976, a similar accelerator was inaugurated at CERN, and named Super Proton Synchrotron (SPS). It inhabited an underground tunnel, 7 kilometres in circumference under the Swiss–French border, part of it in one country and part in the other, so that the accelerated protons had to cross the border twice each revolution. (It was jokingly told that the SPS designers were afraid that Charles De Gaulle, in his efforts to emphasize France's independent stand, might not allow such frequent free crossings of the border . . .)

The 1980s have seen the accelerators breaking into the TeV range (1 TeV = 1000 GeV = 10^{12} eV). Fermilab was the pioneer in this domain, with a new ring installed in the existing synchrotron tunnel, under the old ring, and equipped with powerful superconducting magnets. This accelerator, renamed the Tevatron, began to produce 1000-GeV protons in 1984 (see section 10.8).

Besides pushing up the accessible energies, another very important development in accelerator technology has taken place in the last two decades. We shall say a few words about it here and discuss it more fully in Chapter 8. All the accelerators described so far are based on collisions between moving particles and a stationary target. They are sometimes called 'fixed-target accelerators'. Due to the conservation of linear momentum, only a fraction of the kinetic energy of the incident particles can be converted into the mass of new particles in such collisions. Much more efficient would

Figure 4.6. Inside the tunnel of the Super Proton Synchrotron at CERN. To the left are the focusing and bending magnets which surround the vacuum pipe in which the protons travel. (The photograph was taken before the conversion of the SPS into a $p\bar{p}$ collider, described in section 10.8.) (Courtesy CERN.)

be head-on collisions between particles (preferably a particle and its anti-particle) moving in opposite directions. Machines aimed at producing and studying such events appeared in the 1960s and became the main tool in particle research during the 1970s. They are called *storage rings*, or *colliding-beam accelerators*, or *colliders*.

Since 1972, when the linac of SLAC began to operate as an injector for an electron storage ring, most of the important electron accelerators have been of the collider type. The largest proton synchrotrons in the world began operating as colliders in the 1980s (see Chapter 10).

Let us return for the time being to the fixed-target accelerators. The beam of energetic particles which leaves the accelerator can be aimed at a solid, liquid or gas target. The target is encircled by detectors or located within a detector. The accelerated particles collide with the nucleons in the atoms of the target, and in some of these collisions new particles might be produced while in others the incident particles are scattered intact. The products of all the interactions which arrive at the detectors are recorded for further study. Such experiments can either discover new particles, or tell us

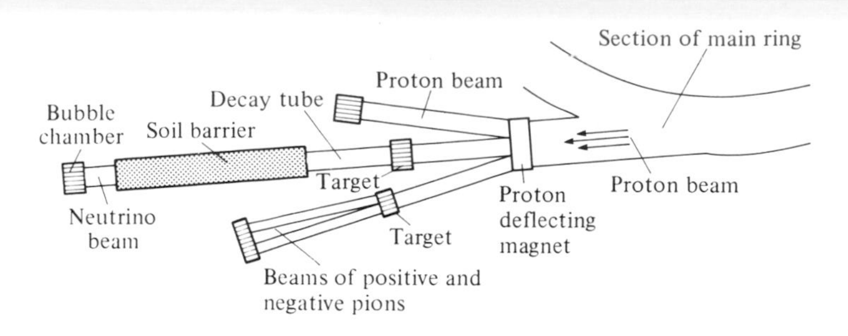

Figure 4.7. Producing neutrino beams in the 500-GeV synchrotron at Fermilab.

something about the interactions between the accelerated particles and the nucleons in the target.

It is often possible to assemble the particles which are produced by the collisions of the primary beam and the target, arrange them in a directed beam, and study the interactions of this 'secondary' beam with another target. For example, the charged pions produced when protons collide with a target have a lifetime of 2.6×10^{-8} seconds, which may seem short in everyday terms, but is actually long enough to enable the gathering of the pions into beams of positive and negative particles, in order to examine the interactions of those beams in a bubble chamber, for example. Similarly, beams of muons, anti-protons and even neutrinos can be produced and handled.

The anti-proton was first observed in 1955 in the Bevatron of Berkeley. By shooting 6-GeV protons at a stationary target, the following reaction was initiated: $p + p \rightarrow p + p + \bar{p} + p$. The physics Nobel prize for 1959 was given for this discovery to Owen Chamberlain and Emilio Segrè (who had been Fermi's assistant in Rome, and was later the head of a research group in Los Alamos during the war).

Fig. 4.7 schematically describes the production and study of neutrino beams at the Fermilab synchrotron, carried out during the 1970s. The protons leave the main ring with an energy of 500 GeV, and are directed to one of three regions, in accordance with the planned experiment. The neutrino region is the central one. After travelling one kilometre in this region, the beam encounters a thick wall of condensed matter. The main products of the collisions between the energetic protons and the nucleons in the material are pions and kaons (particles similar to the pions, but more massive; we shall meet them in the next chapter). By applying magnetic and electric fields, these particles are directed into a 400-metre-long 'decay tube' where part of them decay to muons and neutrinos. A barrier of earth, one kilometre thick, in the end of the decay tube absorbs all the incident particles, except for the neutrinos, which pass through it as if it was not there. The beam emerging from the other side of the barrier is therefore a pure neutrino beam.

On the other side of the barrier there is an immense bubble chamber, containing 30 000 litres of liquid hydrogen, deuterium or neon, which is triggered into operation whenever penetrated by the

Figure 4.8. Installation of a large bubble chamber at CERN. (Courtesy CERN.)

neutrino beam (Fig. 4.8). When an interaction occurs between a neutrino and a nucleon in the chamber, the paths of the produced charged particles appear as tracks of bubbles which are automatically photographed by stereoscopic cameras.

Large electromagnets induce a magnetic field in the chamber and from the curvature of the tracks, the charge and momentum of the particles can be deduced. The bubble chamber allows the interactions of neutrinos and nucleons to be studied much more effectively than Cowan and Reines were able to do in their historical experiment of the 1950s, in which the neutrino was first discovered.

4.3 The physicist as a detective

We owe our knowledge about pions, muons and other elementary particles to research which was carried out over many years by numerous scientists the world over. The following question is inevitable – why spend so much money and put such great effort into studying particles which are so short-lived and which are created

Indeed, the most important facts that we have learned about the nucleon itself – its compositeness, and the nuclear 'glue' – would not have been discovered had we not been interested in the 'strange' particles (see Chapter 5).

only under special circumstances? Since we can build a rather satisfactory model of the atom using just protons, neutrons and electrons, why not confine our studies to these three particles? The answer, in essence, is that it is impossible to understand the physics of some of the particles without knowing something about the whole gang. The strong force which holds together the components of the nucleus cannot be understood in depth without studying other particles affected by the same force, such as the pions. The study of the muon throws light on its family member, the electron, as well. In the eyes of the physicist, stable particles are not more important than those particles which decay in a tiny fraction of a second. After all, stability is not such a basic property. As we have seen, the electron is a stable particle only 'incidentally' and the neutron is not at all stable when outside the nucleus. A millionth of a second, 10^{-8} seconds and even 10^{-16} seconds – the life expectancy of π^0 – seem to us as very short intervals indeed, but these are rather long periods on the sub-atomic scale. (The 'natural' unit of time in this scale is 10^{-23} seconds – the time required for a particle whose speed approaches the speed of light to travel a distance equal to the range of the strong force.)

Particle physicists consider therefore the short-lived particles no less respectable than the stable ones, and they devote intensive efforts to investigate them.

When we look over a table which summarizes the various properties of some short-lived particles, it is hard to realize just how much work has gone into each and every measurement.

Even since the research of particles has been transferred from cosmic rays to accelerators, considerable thought and effort has gone into each experiment. Determination of the properties of a new particle by the examination of its tracks in a cloud chamber or photographic emulsion was backbreaking work, and strewn with errors and reconsiderations, like detective work. Often the theory has been used to help narrow down the various possibilities and to choose the most likely explanation for an experimental result. In some cases, physicists spent years on the wrong track, mistaking or doubting the identification of a certain particle, until enough experimental data could be accumulated to give a definite identification. We cannot, in this book, go into a detailed description of all the many sophisticated experiments which helped physicists discover the secrets of the particle world. However, in order to give the reader some idea of the ways in which the research was carried out, we will review in brief some of the experiments which were carried out during the 1940s and 1950s in order to gather data concerning pions and muons. This section may be skipped or returned to later on, without affecting the continuity. So if you are anxious to read on about exciting discoveries of new particles, feel free to start Chapter 5. However, if you are overwhelmed by curiosity concerning the manner in which the elementary particles were measured and weighed, be patient and continue reading.

Masses and lifetimes

The source of pions and muons in accelerators are protons accelerated to energies of hundreds of MeV, which then strike a stationary target. The collision between a fast-moving proton and a nucleon in the stationary nucleus results, among other things, in the formation of pions. If the pions are allowed to decay, their decay products, namely muons, can be investigated. Early estimates of the mass of the pion and muon were obtained, as you will recall, from an analysis of the tracks of cosmic rays in cloud chamber photographs and in photographic emulsions. In these photographs, however, too many factors were unknown and the results were inexact. More precise measurements were obtained when physicists began to use particle accelerators. In a typical experiment in such an accelerator a proton beam strikes a target, so that some of the protons collide with nuclei and create pions. The beam then continues a certain distance, during which some of the pions decay into muons, and finally enters a magnetic field. The beam is now composed of a mixture of pions, muons and protons. The magnetic field deflects the positive particles in one direction and the negative particles in the opposite direction. The greater the momentum of the particle, the smaller its deviation. The beam is thus scattered in a manner which enables us to separate particles possessing different values of momentum. After the separation, a portion of the beam, composed of particles possessing the same momentum, strikes a photographic emulsion and is brought to a stop. The length of the track which each particle leaves behind in the emulsion is a function of its energy. When we examine the paths left by the positive particles, for example, we find three types of tracks: those of protons, π^+ and μ^+. While all three possess identical momenta, their masses, and thus their energies, differ. Since the momentum is precisely known, one can calculate the mass of the particle from its range in the emulsion (which is a function of its energy) and by comparing the tracks of the protons, pions and muons. From these experiments the masses of the pions and muons were determined to a precision of four digits. The set-up in which the experiments were done is schematically

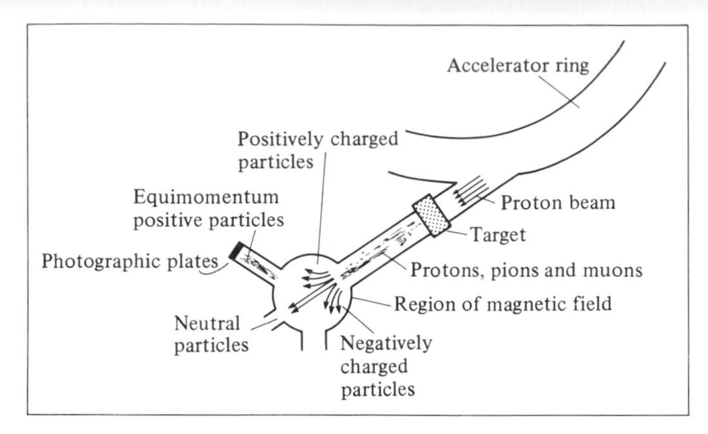

Figure 4.9. Accurate measurements of the masses of the charged muons and pions.

described in Fig. 4.9. The mass of the negatively charged muon was determined with even greater precision by measuring the wavelengths of photons (X-rays) emitted when muons are absorbed by matter and go into orbit around the nucleus, jumping into orbits closer and closer to the nucleus.

The average lifetimes of pions and muons can be determined by examining a beam of known momentum and comparing the relative numbers of particles in the beam at various distances from the source of the beam, or from examination of numerous decay photographs from a bubble chamber. The experimental results were $(2.56 \pm 0.05) \times 10^{-8}$ seconds for charged pions, and $(2.203 \pm 0.004) \times 10^{-6}$ seconds for muons.

We have mentioned that the decay of a pion into a muon is accompanied by neutrino formation. This fact was discovered by examining photographs of such decays in emulsions, and in cloud and bubble chambers. The photographs showed that when a pion was brought to a halt in matter and decayed into a muon, there was always a constant difference between the rest mass of the pion and the energy of the emitted muon (rest mass plus kinetic energy). It was thus evident that an additional particle was produced in the course of this decay. Since the masses of the muon and pion were known, the laws of conservation of energy and momentum enabled the researchers to calculate the mass of the additional particle. Calculations show its mass to be zero. If it were a photon, its energy would be sufficient for the formation of an electron–positron pair. However, in an experiment carried out in 1950 it was found that no such tracks existed adjacent to the point where the decay of the pion took place. Thus, the additional particle must be a neutrino.

Electronic and muonic neutrinos

In the 1950s there was no proof as yet that two types of neutrino did indeed exist, but a few physicists guessed that the neutrino which accompanied muon formation was not identical to that formed together with the electron (in beta decay). This assumption was based on two facts associated with the decay of the muon into an electron.

In addition to these arguments, M. Gell-Mann pointed out in 1960 that more profound reasons, connected with the nature of the basic forces, required the existence of the muonic neutrino.

(*a*) An examination of the energy and momentum of the electron produced in the decay process revealed that *two* neutrino particles are always formed along with it.

(*b*) The decay of a muon into an electron *and a photon* had never been observed ($\mu^- \not\rightarrow e^- + \gamma$ or $\mu^+ \not\rightarrow e^+ + \gamma$) despite the fact that this appeared at first glance to be a possible mode of decay.

These two facts were explained by B. Pontecorvo as follows: apparently, the muon carries a type of charge, as yet unknown, which the electron does not possess. Therefore, its decay into an electron must be accompanied by the formation of another particle which carries this 'muonic charge'. This role is apparently played by one of the two neutrinos which are formed during muon decay, hence known as the 'neutrino of the muon' or the muonic neutrino (ν_μ). The second neutrino appears to be the familiar 'anti-neutrino of the electron', which appears together with the electron in beta decay, as well ($\bar{\nu}_e$). If this explanation is not entirely clear, do not fret, these matters will be explained further in Chapter 7 which deals with conservation laws.

In 1962 an experiment was carried out in the 33-GeV proton synchrotron at Brookhaven in order to determine whether the muonic neutrino and the electronic neutrino are actually different particles. In the experiment, physicists exploited the 'reverse process' of beta decay:

$$\bar{\nu}_e + p \rightarrow n + e^+$$

(with which Cowan and Reines discovered the neutrino) to see whether it could take place with the neutrino formed in the pion-to-muon decay. The experimental set-up is schematically described in Fig. 4.10.

A proton beam, accelerated to an energy of 15 GeV, struck a target of beryllium. A powerful beam of pions (as well as other particles) subsequently emanated from the beryllium. After the beam had travelled a distance of about 20 metres, during which a number of pions decayed into muons and neutrinos, it met an iron wall 13 metres thick. This wall stopped all the particles in the beam,

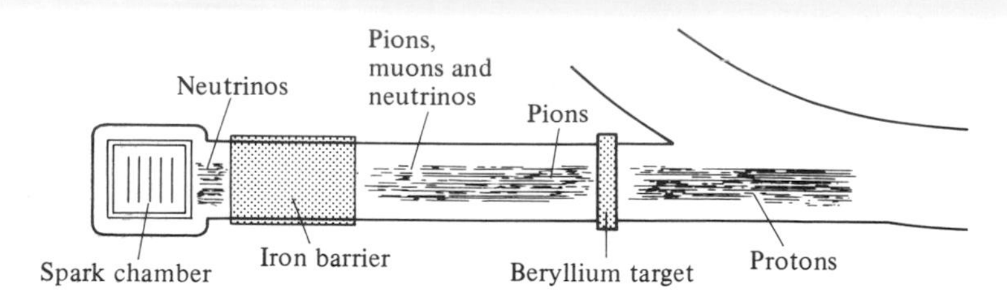

Figure 4.10. The experiment which demonstrated that the muonic neutrino is not the same as the electronic neutrino. A proton beam struck a target of beryllium. Pions subsequently emanated from the beryllium, and some of them decayed into muons and neutrinos. An iron barrier stopped all the particles except the neutrinos. A spark chamber detected the particles which were produced in interactions between neutrinos and nucleons.

with the exception of the neutrinos. The neutrinos passed through the barrier without changing their direction of motion and finally reached a giant spark chamber which contained thick aluminium plates, each with an area of 1.33 × 1.33 square metres. To prevent disturbances caused by cosmic rays, the spark chamber was surrounded by scintillation counters, which ceased its operation every time they discovered that a particle from the outside had penetrated the system. The cameras and the voltage between the plates were activated automatically for extremely short time intervals only when the particle beam was emitted by the accelerator. Every time a neutrino was absorbed by a proton or neutron in the aluminium plate and 'kicked out' a charged particle, a track of sparks was formed. A few dozen events of this kind were recorded in the course of the experiment. An examination of the length of the track and its density revealed that all the tracks were those of muons produced in one of the following processes:

$$\bar{\nu}_\mu + p \rightarrow n + \mu^+$$
$$\nu_\mu + n \rightarrow p + \mu^-$$

The experiment thus proved that 'the neutrino of the muon' produces only muons, and not electrons, and established that it differs from 'the neutrino of the electron'. Thus it was found that at least two types of neutrino and two types of anti-neutrino exist: ν_e, $\bar{\nu}_e$, ν_μ, $\bar{\nu}_\mu$, and that the decay processes of the muon are actually as follows:

$$\mu^- \rightarrow e^- + \bar{\nu}_e + \nu_\mu$$
$$\mu^+ \rightarrow e^+ + \nu_e + \bar{\nu}_\mu$$

These conclusions were later confirmed by further experiments in bubble chambers.

Measuring spins

The spin of new particles may be determined by two methods:

(*a*) Using the law of conservation of angular momentum, the spin of a new particle can be derived from our knowledge of the spins of other particles participating in the process.

(*b*) By examining the angles at which the products are emitted with respect to the directions of the original particles. Theoretical considerations show that the spin has an effect on these angles, and also influences the relationship between the energy of the original particles and the probability of the interaction.

The assumption that the spin of π^+ is zero was first derived from an analysis of the processes $p + p \rightleftarrows \pi^+ + d$ (d represents deuterium). This assumption was proven beyond all doubt by the examination of additional experimental evidence which had accumulated by 1953.

An analysis of the processes in which π^+ and π^- participate showed that the spin of π^- is also zero. From the law of conservation of angular momentum and the examination of processes such as $\pi^+ \rightarrow \mu^+ + \nu_\mu$, $\pi^- \rightarrow \mu^- + \bar{\nu}_\mu$, it stems that the spin of the muon as well as the spin of its neutrino is $\frac{1}{2}$.

The neutral pion

The methods previously described for evaluating masses and lifetimes work fine when applied to charged particles. Neutral particles, however, such as π^0, leave no visible tracks behind them; their paths cannot be detected by cloud and bubble chambers or in photographic emulsions. Therefore, their study must be carried out by indirect means.

The first hint that a neutral pion exists stemmed, as we have already mentioned, from theoretical considerations. The first direct proofs for its existence were provided in 1950. Photons are among the particles formed when a rapidly moving proton beam strikes a target. A study of those photons was carried out in the cyclotron of the University of California at Berkeley. The photons struck a tantalum plate and the electron–positron pairs which were formed were deflected by a magnetic field and detected by counters (ionization chambers) which also recorded the energy of the incident particle. From the energy of the electrons and positrons one could learn about the energy of the photons which had produced them. It was found that their energy did not correspond to any known

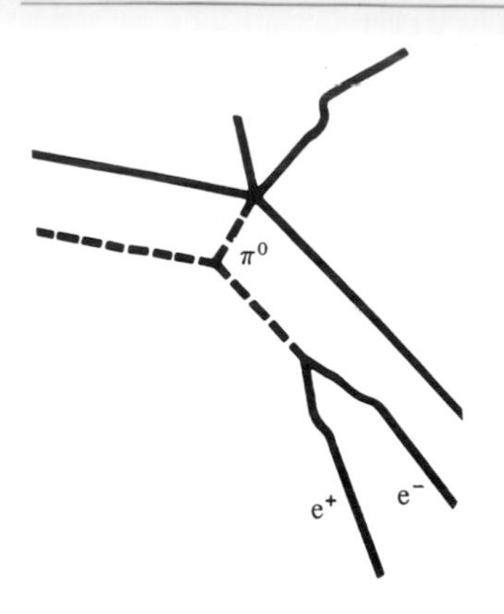

Figure 4.11. Reconstruction of a 'star' formed by a cosmic particle impinging on a photographic plate. Besides the trajectories which emerge from the centre of the star, paths of an e^-e^+ pair appear, which were apparently produced by one of the photons formed in the decay of a neutral pion. The tracks of the π^0 and the photons were not recorded in the photographic emulsion, which is sensitive only to charged particles.

process, but could be elegantly explained if it was assumed that they are formed in the decay of a neutral meson with a mass of about 150 MeV, which decays into two photons.

An even more direct proof was provided in the same year by A. G. Carlson, J. E. Hooper and D. T. King at the University of Bristol. They examined photographic plates which had been elevated to a height of 21 kilometres by balloons. When a rapidly moving proton from cosmic rays strikes such a plate, a great number of paths may be seen to emanate from a single point. In such a pattern, known as a 'star', tracks of the particle π^+ and π^- are recorded. In their vicinity, however, tracks of electron–positron pairs were also observed (see Fig. 4.11). The energy of the photon from which they arose was determined from the analysis of the electron–positron tracks (density of the grains, length, deviation from a straight line). From the same photograph the direction of the photon may be ascertained as well, and it was found that it does not emanate from the centre of the 'star'. It can be accounted for as a decay product of π^0 which was formed together with the charged pions. If the π^0 were brought to a halt before its decay, all the photons should possess identical energies – half the rest mass of the neutral pion. Since, however, it decays while in motion, the photons possessed different values of energy. The mass of π^0 was estimated from an analysis of these energies to be 150 ± 10 MeV. More precise estimates were obtained by examining the collision between π^- and a proton. In some of these cases, π^0 is formed: $\pi^- + p \rightarrow \pi^0 + n$, and an analysis of the results led to the conclusion that the difference between the masses of the pions is:

$$m_{\pi^-} - m_{\pi^0} = (5.4 + 1)\ \text{MeV}$$

It was difficult to determine the lifetime of π^0 from photographs, since the length of its path, from the point of its formation to the point of decay, is too short to be measured accurately. An experiment carried out in 1965 determined its lifetime indirectly, by measuring the probability of π^0 formation when an energetic gamma photon (1000 MeV) collides with a nucleus. Theoretical calculations fixed the spin of π^0 at zero, i.e. the same as that of charged pions. This value has been verified by the millions of π^0 produced in experiments to date.

Chapter Five

Strange particles

5.1 The dawn of a new era

In 1947, after the considerable fuss and confusion surrounding the pion and muon had been resolved, it again seemed as if the physics of elementary particles could rest on its laurels. The number of known particles (including those whose existence was only postulated at the time, but later proven) stood at 14. The four 'pioneers' were the photon, the proton, the neutron and the electron; these were joined by three anti-particles (e^+, $\bar{n}$, $\bar{p}^-$) as well as the neutrino–anti-neutrino pair (physicists still did not distinguish between ν_e and ν_μ), three pions and two muons. Aside from the riddle presented by the muon, the 'roles' of all the other particles in the physicist's world picture were apparently clear: the electron and the nucleons make up the atom. The photon carries the electromagnetic force, and the pion, the strong force. Neutrinos are necessary to maintain the laws of conservation in certain decay processes such as beta decay. The existence of anti-particles was understood as the result of some basic symmetry of nature predicted by Dirac's equation. And yet, just as in 1932, when all the riddles seemed to have been answered another revolution was just around the corner. In 1947, strange unrecognized tracks of cosmic particles were recorded in cloud chambers; these tracks had not been previously observed, and demanded an explanation. After considerable research effort these tracks were deciphered and within 10 years the number of elementary particles more than doubled. New conservation laws were discovered and the basis for classification of particles into families was established. Nor did the revolution turn its back on research methods. A new generation of gigantic accelerators was constructed and the emphasis of particle research passed from cosmic rays to these accelerators. The bubble chamber, invented in 1952, replaced the cloud chamber as the major research apparatus in accelerators, and allowed more efficient examinations of new particles.

Unknown tracks

The first tracks of unknown cosmic particles appeared in cloud chambers in 1944, but the first clear results were presented by G. D. Rochester and C. C. Butler from Manchester University in 1947. In cloud chamber photographs at sea level, there appeared a pair of tracks, emanating from a single point, in the shape of the letter V, which were deflected in opposite directions by the application of a magnetic field. An analysis of the photograph proved that an unknown neutral particle, possessing a mass of about 500 MeV (between the mass of a proton and that of a pion) disintegrated in the cloud chamber to π^+ and π^-. A broken track in a second photograph was identified as the decay of a charged particle possessing a mass of about 500 MeV into a pair of pions, one neutral, the other charged.

Researchers all over the world swooped down on photographs of cosmic rays, taken in cloud chambers at various elevations, and sought additional evidence for similar particles. The group of C. F. Powell from Bristol, which specialized in superior photographic emulsions, joined the hunt in 1949. Their quest was successful. It was shown that when a primary cosmic particle possessed energy in excess of one GeV (10^9 eV), its interaction with a nucleus in the atmosphere could produce various new particles which were heavier than the familiar pion. Due to the rarity of such an occurrence, tens or hundreds of photographs of cosmic radiation had to be studied in order to come across finally a good photograph of one of the new particles. This explains why these particles had not been discovered earlier, along with the pion and the muon.

The tracks, found in cloud chamber photographs and photographic emulsions, provided evidence of the decay of neutral particles possessing masses of about 500 MeV as well as particles possessing about twice this mass. In addition, tracks of positive and negative particles possessing similar masses were also discovered. During the years 1947–53, cosmic rays provided the sole source for these particles, and since the researchers could control neither the energy nor the direction of cosmic rays, systematic study of the new particles was extremely difficult. It sometimes happened that a single particle was given a number of names, or that a number of different particles were initially given the same name. Powell's group, for example, found, on film, the track of a positive particle which decayed into three pions, and named it the tau meson, $\tau^+ (\tau^+ \rightarrow \pi^+\pi^+\pi^-)$. Another positive particle with the same mass (within the precision of the measurement) decayed into two pions. It was named the theta meson, $\theta^+ (\theta^+ \rightarrow \pi^+\pi^0)$. Because of a conservation law known as 'the law of conservation of parity' (see Chapter 7), physicists were convinced that τ and θ were two distinct particles with different properties. Only in 1957 was it finally determined that these are one and the same particle, which can decay in a great number of ways. This particle is known today as the K meson (or the kaon).

5.2 Strange particles

In 1952, the 'Cosmotron' accelerator was inaugurated at the Brookhaven National Laboratory in the state of New York (see section 4.2). Since this accelerator was capable of accelerating protons to energies in the GeV range, it could produce a wide variety of new particles possessing predetermined energies and in desired directions, facilitating the investigation of their properties. Within a few years other accelerators in the GeV range were constructed in the USA and the properties of the new particles became clearer. It became evident that the new particles could be divided into two groups. The first group consisted of five particles possessing masses of about 500 MeV, which are similar in certain respects to pions (they possess zero spin, and are affected by the strong force) and may also be considered mesons (an accurate definition of mesons will appear later). In this group we find a quartet of similar particles known as K mesons, or kaons, one positive, one negative and two neutral, as well as a single neutral particle, the eta nought (η^0). The positive and negative kaons are a particle–anti-particle pair, as are the two neutral kaons. The η^0 is its own anti-particle. The important properties of these particles are shown in Table 5.1. (The neutral kaons have two decay modes: a fast one and a slow one. This phenomenon will be discussed in Chapter 7.)

In the second group we find new particles possessing half-integer spins ($\frac{1}{2}$ or $\frac{3}{2}$) whose masses are greater than those of the nucleons; nucleons are, in fact, among their decay products. These particles were named hyperons. The hyperon group consists of seven particles: a single neutral particle known as the lambda (Λ^0); a trio of particles possessing charges of +1, 0, −1, called sigma ($\Sigma^+, \Sigma^0, \Sigma^-$); a pair of particles possessing charges of −1 and 0, known as xi (Ξ^0, Ξ^-). In 1964 they were joined by a solitary negative particle, the

Table 5.1. *Mesons of spin zero known in the 1960s*

Particle	Mass (MeV)	Average lifetime (seconds)	Main decay modes: Products	Fraction (%)
$\pi^{\pm}$	139.6	2.6×10^{-8}	$\mu\nu$	~100
π^0	135.0	0.8×10^{-6}	$\gamma\gamma$	98.8
			γe^+e^-	1.2
$K^{\pm}$	493.7	1.2×10^{-8}	$\mu\nu$	63.5
			$\pi^{\pm}\pi^0$	21
			$\pi^{\pm}\pi^+\pi^-$	5.5
			$\pi^{\pm}\pi^0\pi^0$	1.7
			$\mu^{\pm}\pi^0\nu$	3.2
			$e^{\pm}\pi^0\nu$	4.8
$\bar{K}^0, K^0$	497.7	$(K^0_S)8.9\times10^{-11}$	$\pi^+\pi^-$	68.6
			$\pi^0\pi^0$	31.4
		$(K^0_L)5.2\times10^{-8}$	$\pi^0\pi^0\pi^0$	21.5
			$\pi^+\pi^-\pi^0$	12.4
			$\pi^{\pm}\mu^{\mp}\nu$	27
			$\pi^{\pm}e^{\mp}\nu$	38.8
η^0	548.8	2.5×10^{-19}	$\gamma\gamma$	38
			$\pi^0\gamma\gamma$	3
			$\pi^0\pi^0\pi^0$	30
			$\pi^+\pi^-\pi^0$	23.5
			$\pi^+\pi^-\gamma$	5

omega minus (Ω^-). Each of these particles has its own anti-particle. Σ^- and Σ^+ are not particle and anti-particle, as are K^- and K^+! The anti-particle of Σ^- is a positive particle which we will denote $\bar{\Sigma}^+$ (sometimes this particle is represented by $\overline{\Sigma^-}$, but we prefer the symbol $\bar{\Sigma}^+$, since the sign in this case is indicative of the actual charge of the particle). The anti-particle of Σ^+ is $\bar{\Sigma}^-$, that of Σ^0 is $\bar{\Sigma}^0$, and so on. The Ω^- has an anti-particle too – the positive $\bar{\Omega}^+$. The important properties of the hyperons are summarized in Table 5.2. This table includes our old acquaintances, the proton and neutron, as well. All the particles in this table possess spin $\frac{1}{2}$, except the Ω^- which has a spin of $\frac{3}{2}$.

Most of the particles appearing in Tables 5.1 and 5.2 decay within 10^{-10}–10^{-8} seconds of their formation, while the lifetimes of some of them are even shorter (up to 10^{-20} seconds). Most were discovered during the 1950s, but the η^0 and Ω^- – with shorter lifetimes or greater masses than the others – were not discovered until the 1960s. Their discovery provides a good example of theoretical prediction, similar to the prediction of the meson by Yukawa: they were discovered experimentally after theoretical physicists had successfully predicted not only their existence but their properties as well.

Table 5.2. *Nucleons and hyperons*

Particle	Mass (MeV)	Average lifetime (seconds)	Main decay modes: Products	Main decay modes: Fraction (%)
p	938.3	Stable		
n	939.6	920	$pe^-\nu$	100
Λ^0	1115.6	2.6×10^{-10}	$p\pi^-$	65
			$n\pi^0$	35
Σ^+	1189.4	0.8×10^{-10}	$p\pi^0$	51.6
			$n\pi^+$	48.4
Σ^0	1192.5	5.8×10^{-20}	$\Lambda^0\gamma$	100
Σ^-	1197.3	1.5×10^{-10}	$n\pi^-$	100
Ξ^0	1314.9	3×10^{-10}	$\Lambda^0\pi^0$	100
Ξ^-	1321.3	1.6×10^{-10}	$\Lambda^0\pi^-$	100
Ω^-	1672.2	0.8×10^{-10}	$\Xi^0\pi^-$	23
			$\Xi^-\pi^0$	8
			Λ^0K^-	69

We will read more in later chapters about these scientific successes which advanced our understanding of the world of particles.

It is interesting to note that when the kaon and hyperons were first discovered, they were given the name 'strange particles', because they did not fit into the solid picture of the world which the physicists had constructed, and because their study revealed several puzzles. For example, in experiments using accelerators it was found that when these particles were formed in proton–proton or pion–proton collisions, they were always produced in pairs. This raised the suspicion that their formation is affected by an unknown conservation law, just as the electron–positron pair formation from the photon is an indication of the law of charge conservation. Another cause for wonder is related to their lifetimes: according to the theory, these particles should have decayed much more rapidly than they actually do – within 10^{-23} seconds, and not within 10^{-10} or 10^{-8} seconds as observed experimentally. (Later you will understand why.) An hypothesis published by two young physicists in 1953 clearly accounted for all these 'strange' properties.

M. Gell-Mann in the USA, and K. Nishijima in Japan independently arrived at the same idea (Gell-Mann was then 23 years old and Nishijima 26). According to the theory, these new particles share a common property which the 'veteran' particles lack. The traditional humour of physicists in name giving came into play when Gell-Mann named this property 'strangeness' to commemorate the fact that the particles bearing this property once appeared 'strange'.

(Another example of this sort of humour is the name given by physicists to denote the unit of area equal to 10^{-24} square centimetres – a barn!) We will deal with the property of 'strangeness' in greater detail in the next chapter.

The first hint of the existence of strange particles was the V tracks which were discovered towards the end of the 1940s. Discovery of most of the particles which appear in Tables 5.1 and 5.2 was completed by the early 1960s. Thus the investigation was carried out over the course of about 10 years. The Λ^0 is responsible for some of the V tracks which were discovered in cosmic ray photographs towards the end of the 1940s (the rest were produced by kaons), and thus it was the first hyperon to be discovered. Σ^+ and Σ^- were also first discovered in cosmic rays. Their tracks could be observed as emanating from the 'star' of pion tracks, and were characterized by their much greater length, as compared to the pion tracks. In the accelerator, they can be formed in the collision between kaon and proton:

$$K^- + p \rightarrow \begin{cases} \Sigma^+ + \pi^- \\ \Sigma^- + \pi^+ \end{cases}$$

The lifetime of Σ^- is twice as long as that of Σ^+, since the latter has two channels of decay, whereas Σ^- only has one. (See table 5.2).

The model of Gell-Mann and Nishijima predicted (later we shall understand how) that the sigma particles constitute a trio, and so there must exist a Σ^0 which decays as follows: $\Sigma^0 \rightarrow \Lambda^0 + \gamma$. Shortly after that 'prediction' the Σ^0 was discovered experimentally by its decay products.

The negative xi (Ξ^-) was first discovered in cloud chamber photographs, decaying as follows: $\Xi^- \rightarrow \Lambda^0 + \pi^-$. The model of Gell-Mann and Nishijima predicted that a neutral xi also exists, and later it was indeed discovered experimentally.

The omega minus is the heaviest hyperon. It was discovered in 1964 in the bubble chamber at Brookhaven, following independent predictions in 1962 of its existence and its properties by Yuval Ne'eman of Tel Aviv and Murray Gell-Mann in the USA. The theoretical predictions set the mass of Ω^- at 1675 MeV (a deviation of only 0.2 per cent from the measured value) and correctly predicted its manner of formation and its modes of decay (we will come back to this subject later).

The theory of Gell-Mann and Ne'eman also predicted the existence of η^0, but because of its short lifetime it could not be observed in a bubble chamber or emulsion: its track is too short, even when it moves at speeds approaching the speed of light. The η^0 was finally discovered by its decay products. It was found that in the collision between π^+ and a stationary proton, at energies of 1100 MeV, three pions (π^-, π^0, π^+) are formed, in addition to the two colliding particles. The total energy and momentum of the three pions is always the same. This phenomenon cannot be explained unless it is assumed that the initial product of the collision is one particle – the η^0 – and it then decays into the three pions.

Many surprises awaited physicists who researched the properties of the strange particles – they encountered several phenomena which were unknown when the number of known particles was smaller. We have mentioned the θ–τ riddle which was solved by proving that the two particles are none other than a charged kaon (as mentioned before we will discuss this 'riddle' in greater detail later). The investigation into the properties of the neutral kaon also had some surprises in store; we will discuss these after we become better acquainted with the concept of 'strangeness'.

Chapter Six

Basic forces and the classification of particles

Biologists who study the animal or vegetable kingdoms make use of a classification system which breaks down the hundreds of thousands of species into various groups, classes and families, according to generally accepted criteria. Were it not for the elucidation of this classification system, by Carl von Linné (Linnaeus) in the eighteenth century, Darwin would not have been able to conceive of the theory of evolution and biologists would have encountered great difficulties in doing research and in communicating their findings to their colleagues. While the number of various particles we have encountered so far is by no means as overwhelming as the number of living species, the overall picture at this stage may seem a little confusing to the reader who is making his first acquaintance with the world of elementary particles. Undoubtedly, the physicists who began discovering particle after particle felt the same way. A system of classification was thus essential in particle physics, as well. When we sort the particles into various categories, not only does the general picture become clearer, but the physical laws which reign over the jungle of particles also become increasingly apparent. In previous chapters we occasionally hinted that one group of particles or another constitute a single family. However, if our classification is to have a solid base, we must make use of clearcut, precise definitions. This is one of the purposes of the present chapter. Since an important criterion for our classification will be the type of forces which affect a given particle, we will start off our chapter by taking a look at the four basic forces, or 'interactions', as they are called by the physicists. After we have sorted the particles into various groups, we will discuss in the next chapter the conservation laws which have a great bearing on the world of elementary particles.

6.1 The four basic forces

All the forces in the universe stem from four basic forces:

(1) the force of gravity;
(2) the electromagnetic force;
(3) the weak force;
(4) the strong force.

As we shall soon see, the electromagnetic and weak forces are now regarded as two manifestations of one basic force.

These forces cause all the variety, change and beauty in the universe. Without them there would be neither attraction nor repulsion. Bodies would not collide nor have any effect whatsoever on one another – they would just pass right through each other. All four forces have been mentioned in previous chapters. In this section we will deal with them in greater detail, hoping to throw more light on their properties.

In classical physics, a force is defined as something capable of either affecting the motion of a body (for example arresting its motion, speeding it up or slowing it down, changing its direction and so on) or distorting its shape. In particle physics force is the cause of every change, reaction, creation and disintegration. When two particles collide, resulting in the creation of new particles, a certain force is responsible. When a particle decays spontaneously into other particles, the force responsible may be seen as acting between the original particle and its own decay products (despite the fact that the products do not exist prior to the decay of the original particle, and when they do appear, the original particle does not exist any more).

Since the roles played by the forces in particle physics are so different from those traditionally attributed to them, physicists prefer to refer to them as *the four basic interactions* rather than the four basic forces.

An interaction may be described as a mutual action or influence between two or more particles. The decay of a particle into other particles is also considered to be an interaction, since, as we have mentioned, such a process may be regarded as an event involving the original particle and its products. It was found that any process in which particles are created or destroyed may be looked upon, at least approximately, as a result of the action of *a single basic force*. This is because the intensities of the various forces are quite different, as we shall see in the next section. If, for example, a collision between the particles A and B results in the creation of the particles C and D: $A + B \rightarrow C + D$; and if the process can take place due to the action of the strong force, then the much weaker electromagnetic force will not affect the process substantially, and we can classify that process as a strong interaction. Similarly, the spontaneous decay of a particle can also be considered as a process induced by a single basic force.

We will learn that among the various particles there are those that are influenced by all the forces, and others that are affected by only some of them. In order for a certain process between two particles to take place, both must be capable of being influenced by the force responsible for that process. For example, a process in which an electron takes part cannot be a strong interaction, since the strong force does not act on electrons. An additional condition is that the interaction is not in contradiction to certain conservation laws which we will read about later on (Chapter 7). We will now discuss the four basic forces, one by one.

The force of gravity

Gravitation, or the force of gravity, was the first force to be identified by physicists. The law of gravitation, which was formulated by Newton in the 1680s, states that every piece of matter in the universe attracts every other piece of matter with a force that is proportional to the product of their masses and inversely proportional to the square of the distance between them. In mathematical form:

$$F = G\frac{m_1m_2}{r^2}$$

where F is the gravitational force measured in newtons, m_1 and m_2 are the two masses (in kilograms), and r is the distance (in metres). G is known as the gravitational constant, its value being:

$$G = 6.67 \times 10^{-11}\ \mathrm{N \cdot m^2/kg^2}$$

When we calculate the gravitational force between two large bodies, r is measured as the distance between two points which are called the centres of mass or centres of gravity. For example, a spherical mass of uniform density applies a gravitational force equivalent to that which would be applied were all the mass concentrated at the centre of the sphere.

Legend has it that gravitation 'struck' Newton in the form of an apple falling from a tree. We do not know whether there is any truth to this legend, but we can be sure that Newton did formulate his laws after carefully studying Kepler's observations of planetary motion and Galileo Galilei's experiments with falling objects. Newton himself said, in humility that should be an example to follow, that he managed to see to a great distance because he stood on giants' shoulders.

The first experimental verification of the general theory of relativity was obtained, as already mentioned, during a solar eclipse which occurred on 29 March 1919. Two expeditions which had stationed themselves in places where the eclipse would be total – in northern Brazil and on Principe Island, off the coast of Spanish Guinea in West Africa – photographed the stars near the darkened circle of the sun. The photographs were later compared with photographs of the same region of the sky, taken five months earlier at the Greenwich Observatory, and the comparison showed that the light had indeed been deflected by 1.6 seconds of arc as it passed near the sun. The expedition to Africa had been organized and led by the renowned British astronomer and astrophysicist, Sir Arthur Eddington, who was known for his research on the internal structure of the stars and was an enthusiastic supporter of the theory of relativity from the very beginning, before it became widely accepted.

At a conference held in Jerusalem in March 1979, to mark the hundredth anniversary of the birth of Einstein, P. A. M. Dirac. who had been a student of Eddington 60 years earlier, related how the mission had been planned during World War One, and how Eddington had been plagued with anxiety lest the war would prevent its implementation, causing a rare opportunity to be missed. As it turned out, the war ended in time and Eddington breathed a sigh of relief.

Albert Einstein, in his theory of general relativity, demonstrated that while Newton's law of gravitation applies perfectly to masses and distances which are static and not too great, it is actually part of a more general and more complex law. One of the conclusions of the theory of general relativity is that the motion of the planets revolving around the sun deviates slightly from that predicted by Newtonian mechanics. Such a deviation was indeed measured in the orbit of the planet Mercury. Another interesting conclusion is that the force of gravity has an effect on photons despite the fact that they have no mass at all. This conclusion was verified in an impressive manner during the full solar eclipse of 1919 and in many additional experiments that have been carried out since then. It became apparent that light rays (or radio waves and the like) are attracted by massive bodies such as the sun, and are deviated from their straight paths when they pass near them.

Although gravity acts on all particles, its importance in experimental particle physics is negligible. The reason is simple – the force of gravity is extremely feeble when compared with the three other interactions at the present energies obtained in accelerators and its effect on the particles known to us can be ignored. Not one observed process of creation or decay of particles is due to this force.

It might, indeed, seem ludicrous to call the only force which we are constantly aware of 'a feeble force'. After all, isn't it the force of gravity which prevents us from floating about in the air, gives us palpitations when a piece of glass or china slips from our hands and holds the planets in their orbits? But we are so very conscious of the force of gravity because of the immense mass to which we are attracted – the earth. On the other hand, it is very difficult for us to measure the gravitational attraction between two lead weights (although Sir Henry Cavendish actually did so in a clever experiment which he carried out in 1798). There are reasons for the fact that we are not directly aware of the three other interactions (which are all stronger than gravity) in our daily lives: the ranges of both the strong and weak nuclear forces are extremely small, and electrical forces have no effect on neutral bodies, i.e. bodies whose net electric charge is zero. If just one-thousandth of one per cent of the approximately 10^{30} electrons in our body were transferred to the ground below us, each one of us would be subjected to an electrical attraction equivalent to the weight of 10^{18} tonnes! We shall, in the following pages, make a comparison between the various basic forces *under identical conditions*, and we shall see just how negligible the force of gravity is in comparison with the other forces. For this reason, physicists do not take gravity into account when they consider the various basic interactions which influence particles at

It is interesting that in his later years Eddington had some original ideas that were not always accepted in theoretical physics and the philosophy of physics. He was so taken with the model which postulated that the nucleus was made of electrons and protons, that he refused to abandon it, even after the discovery of the neutron. He insisted on regarding the neutron as a bound state of a proton and electron, and developed a complicated theory to support this notion.

present energies. Nevertheless, any general comprehensive theory of particles and forces must take gravity into account and explain its origin and its relationship to other forces. It should also predict its action at the quantum level, which is still a mystery. Indeed these issues have become very important in the evolution of particle theory in the 1970s and 1980s, and we shall touch upon them in Chapter 10.

We have seen that the action of electrostatic force may be considered as an exchange of photons, and the strong interaction as an exchange of mesons. Is there a similar particle which carries the force of gravity? Physicists think there is. General considerations have led them to the conclusion that the force of gravity is carried by a massless particle of spin 2, which is affected solely by the force of gravity. This particle has been named the graviton. The existence of the graviton has not yet been directly confirmed by experiment. This is not surprising. The fact that the graviton is only affected by gravity greatly decreases the probability of interactions between it and other particles. The 'weakness' of the force of gravity is demonstrated by the fact that the total power of the 'gravitational waves' emitted by the earth while circulating around the sun is only 10 watts.

Finally we should mention one reservation concerning the relative weakness of the force of gravity. Because of the differences in the characteristics of forces, circumstances may arise in which the force of gravity will surpass the electromagnetic force and even the strong force. Such a situation will never take place when a small number of particles are involved, but might arise when a considerable mass is concentrated into a small lump. A massive star whose heat has dissipated collapses under its own weight and condenses due to the influence of gravity. Sometimes this process is so powerful that the atoms of the star are squashed and the electrons are forced into the nucleus where they convert the protons into neutrons (emitting neutrinos). This process manifests the victory for gravity over electromagnetic force. This is how a neutron star, which is composed of closely packed neutrons and whose density is many tonnes per cubic centimetre, is created. The neutrons are bound as if the entire star is one gigantic nucleus (about 10 kilometres in radius). The observed pulsars are deemed to be neutron stars surrounded by clouds of gas. If the mass of the star is big enough, the process does not stop here. Gravity continues to condense the star with a strength which even the repulsion, arising between the neutrons when they are pressed against one another, cannot withstand any more. According to the calculations, the result is a further contraction (gravitational collapse). In this manner the so-called

Black holes are described by equations of Einstein's general relativity, first solved in 1916 by K. Schwarzchild. In the 1970s, Stephen W Hawking of Cambridge University, developing an idea of J. Bekenstein of Ben-Gurion University (Israel), showed that microscopic black holes may have been formed in abundance in the early cosmos. Hawking, who despite a severe amyotrophic lateral sclerosis which confines him to a wheelchair has made important contributions to theoretical astrophysics, later showed that, contrary to the accepted view, black holes do emit radiation and particles due to quantum effects.

The first celestial object to be interpreted as a black hole is located in the Cygnus X-1 binary system. A second one was discovered in 1982 in our nearest neighbour galaxy, the Large Magellanic Cloud.

'black holes' are created. Astronomers believe that they have found evidence for the existence of a number of these celestial bodies.

Theory also predicts that at extremely high energies – such as those that existed in the original Big Bang, some 10^9 years ago – gravity is dominant, and its carrier (the graviton) can produce particle–anti-particle pairs, just as the photon does. Indeed all the matter in the universe may have been created out of the decay of extremely energetic gravitons.

The electromagnetic force

About a century after the formulation of the law of gravitation, the French physicist, Charles A. Coulomb (after whom the unit of electric charge is named), successfully measured the force acting between two stationary electric charges in a vacuum (the electrostatic force) and developed an equation very similar to the one found for gravity:

$$F = k\frac{q_1 q_2}{r^2}$$

F is the force between the charges (in newtons); q_1 and q_2 represent the respective values of the charge (in coulombs), r is the distance separating the two charges (in metres), k is a constant of proportionality – its value being

$$k = 8.987\ \mathrm{N \cdot m^2/C^2}$$

As opposed to the force of gravity, which is always an attractive force, the electrostatic force can either attract (when one of the charges is positive and the other negative) or repel (when both are either positive or negative). The other manifestation of the electromagnetic force is the magnetic force which arises between *moving* charges, in addition to the electrostatic force between them. Both these forces stem from the same basic phenomena. This became clear in the nineteenth century, when James Clerk Maxwell published his theory in which the electric and magnetic fields appear side by side in the same equations. Later it was demonstrated that by using special relativity the magnetic force between moving charges can be presented as a direct consequence of the electrostatic force. Thus we usually speak of a unified electromagnetic force.

Almost all the forces which affect our everyday life, such as the attraction between atoms or molecules, and that between electrons and the nucleus of the atom, as well as all the forces which act within the living organism, stem in some way from the electromagnetic force. This force is also responsible for the existence of all the various kinds of electromagnetic radiation, ranging from radio waves to gamma rays. Any change in the velocity of a charged particle brings about the emission of a photon which, as we have already seen, is the particle which transfers electromagnetic force between bodies.

It is important to note that although the electrostatic force acts only between charged particles, electromagnetic interactions can in general have an effect on uncharged particles as well. The neutrally charged neutron, for example, has a non-zero magnetic moment and is thus influenced by magnetic fields. The photon and the π^0 possess neither charge nor magnetic moment, and yet electromagnetic interactions are responsible for the creation of the photon, its absorption by the atom and its turning into the pair e^+, e^-, as well as for the decay of π^0 into two photons. Actually, in one way or another, this interaction directly affects all particles except for the neutrino and graviton.

The weak force and the electroweak theory

The weak interaction is responsible for a number of phenomena, such as the conversion of a neutron into a proton or vice versa (beta decay), the decay of a pion into a muon and the decay of a muon into an electron. In all the above-mentioned interactions a neutrino is also emitted (neutrinos are the only known particles affected solely by the weak force), and it was once thought that this emission was the 'trademark' of all weak interactions. As it turned out, however, neutrinos do not take part in all weak interactions; for example – in many of the decays of the 'strange' particles, neutrinos are not involved. However, since the late 1930s the analysis of all the processes which are considered as weak interactions brought out the similarities between them and showed that they are definitely of common origin.

One thing common to all weak interactions is that the probability of their occurrence is quite low, indicating the relative weakness of the weak force compared to the strong and electromagnetic forces. The range of the weak force is smaller by at least two orders of magnitude from that of the short-ranged strong force. In order for a weak interaction to occur, the distance between participating particles should be less than 10^{-15} cm.

The theory of the weak force has experienced several alterations and developments since Fermi's pioneering work in the 1930s. In 1958 the American R. E. Marshak and the Indian E. C. G. Sudarshan pointed out that weak processes appeared to involve an action between two currents, resembling the attraction or repulsion between two wires carrying electric currents. R. Feynman and M. Gell-Mann further noted that the currents were partially conserved. Back in 1955, the Soviet physicists S. S. Gerstein and I. B. Zeldovich had already noted that the near equality between the strength of the 'weak' force in neutron beta decay and in muon decay resembles the equality between the electric charge of a proton and that of a positron. Thus, the neutron and the muon seem to carry the same 'weak' charge. The similarity between weak and electromagnetic currents was an important clue. In the 1960s and 1970s a new unifying theory emerged which incorporated the weak and electromagnetic interactions in a single framework as Maxwell's theory unified the electrostatic and magnetic forces. This electroweak theory is sometimes called the Weinberg–Salam theory after the American Steven Weinberg, and the Pakistani Abdus Salam of Imperial College, London, and the International Centre for Theoretical Physics at Trieste (Italy), who independently published the principles of this theory in 1967 and 1968.

Other physicists contributed to the development of the theory as well, among them the American Julian Schwinger, who had already in 1957 pointed out qualitatively a possible way to unify the two interactions, and his student Sheldon Glashow, who further pursued this idea in his doctoral thesis and went on evolving it for years. Another work, deserving special credit, was that of the 24-year-old Gerald 't Hooft, a Dutch research student from Utrecht University, who in 1971 managed to overcome some mathematical hurdles which had blocked the way to the final acceptance of the theory.

According to the Weinberg–Salam theory, the electroweak interactions are carried by four bosons – one positive, one negative and two neutral. In the region of high energies (which might be reached in the accelerators of the 1980s) these particles would be similar to each other. But in the region of moderate energies, like those available in the present accelerators, the symmetry between the particles is broken in such a way that three of them are provided with substantial masses, while the fourth remains with no mass at all. The massless one is of course our old acquaintance, the photon, which furnishes the electromagnetic branch of the unified force with an infinite range. The other three particles, which were designated W^+, W^- and Z^0, should have, according to the theory, masses in the range of 80–100 GeV, and therefore the weak force which they mediate has such a short range. Thus, in spite of the common source of the electromagnetic and weak forces, they differ in their properties in practical cases, and the processes they induce can still be classified separately as weak or electromagnetic interactions.

The best experimental evidence for the new theory would have been to detect those three heavy carriers of the weak force (which were called 'intermediate vector bosons' or 'weakons'). However, when the theory was proposed, no existing accelerator could even approach the required energy range. But the theory had other, less direct, corroborations. The strongest one was the discovery in 1973 of certain processes ('neutral weak currents', see section 10.1) at CERN and Fermilab. In 1979 Salam, Weinberg and Glashow shared the Nobel prize for the electroweak theory even though the weak force carriers had not been detected yet, and their theory thus lacked the decisive experimental proof. Several years later, towards the end of 1982, the circle was completed when the first signs of the W^+, W^- and Z^0 showed up at CERN, with the right masses and other properties as predicted by the theory (see section 10.8). Gerald 't Hooft shared the 1982 physics Wolf prize in Jerusalem with two of the pioneers of quantum electrodynamics, V. Weisskopf and F. Dyson.

The strong force

We mentioned earlier that the strong force is a short-range attractive force between nucleons, which does not depend on electric charge. It is this force which is responsible for nuclear stability, for the phenomenon of nuclear fusion (i.e. the making of helium nuclei from protons and neutrons) and for the creation of particles in collisions between two nucleons and between a nucleon and a meson. The strong force affects all the particles appearing in Tables 5.1 and 5.2, but it does not influence the electron, the muon, the neutrinos, and the carriers of the forces: the $W^{\pm}$, Z^0, photon and graviton.

Yukawa's equation for the intensity of the strong force between nucleons in the nucleus (section 3.4) is only a rough and quite unsatisfactory approximation. First of all, in contrast with electrostatic and gravitational forces, it is not clear whether the strong force between two nucleons is along the shortest line between them. In addition, the strong force between two nucleons is influenced by the presence of other nucleons in the vicinity. For example – the energy required to separate the proton from the neutron in the deuterium nucleus is 2 MeV, but the energy needed to separate the helium nucleus into its four components is 30 MeV! The strong force depends on the direction of spin, and Yukawa's equation does not take this into account. The force attracting two nucleons of opposite spins is weaker than that attracting two nucleons whose spins are in the same direction.

Yukawa's equation does not reflect the fact that the strong interaction is an attractive force only when the distance between the centres of the two nucleons is greater than 1.2×10^{-13} centimetres, and when we try to reduce this distance further we encounter a strong repulsive force, known as the 'hard core repulsion'.

The physicists are now convinced that the understanding of the strong force between the nucleons is attainable only through the understanding of the force acting *inside* the proton and neutron, between their basic constituents. This force will be explained later, when we deal with the quarks (section 9.7).

We have mentioned the electroweak theory which unified the electromagnetic and weak forces. Since the late 1970s, attempts have been made to achieve a greater synthesis by unifying these two forces with the strong force in one coherent theory, or even to find the hidden relationship between all four basic forces. Some of these attempts have indicated that the proton may not be totally stable, but rather have a finite, very long, lifetime. We shall discuss this startling possibility later, when dealing with conservation laws. It is worthwhile to mention that in his last 20 years, Einstein devoted much effort to attempts to find a unified theory of gravity and electromagnetism. Today we know that his attempts were destined to fail, since at that time the other basic forces, which are essential to any such unification, were not yet known.

6.2 The relative strength of the basic forces

Because of the differences in the properties of the four forces, it is as hard to compare them under equivalent conditions, as it would be to compare the level of a judo expert with that of a boxing champion. Let's assume that we want to make a comparison between the electrostatic force and the force of gravity. A simple calculation shows that the electrostatic attraction between the electron and the proton in a hydrogen atom is about 10^{39} times greater than the gravitational attraction between them. Between two protons, however, the ratio of the forces is 10^{36}, and between an electron and a positron, approximately 10^{42}. In all three instances the electrostatic force is identical, while the gravitational force which is mass-dependent varies.

To separate the proton from the neutron in the deuterium nucleus, held together by the strong interaction, would require an energy of 2 MeV, while the separation of two atoms in a molecule held together by electrostatic force requires an energy output of the order of 2 eV. The difference between these figures does not, however, reflect the ratio between the strengths of nuclear and electrostatic force, since the distance between the two particles in the first case is 10^{-13} cm and in the second case, approximately 10^{-8} cm.

Two adjacent protons cannot form a stable structure. Does this mean that their respective electrostatic repulsion is greater than the strong force which attracts them? Not necessarily. We must remember that we are talking about two fermions. Pauli's principle thus applies here, and the two particles must have opposite spins; we have already mentioned that the nuclear force between protons with opposite spins is relatively weak.

The above examples are indicative of the problems which we face when trying to compare various forces. Even so, one of the feats of theoretical physics was to formulate the action of the four forces, using dimensionless factors known as *coupling constants*, whose magnitude is a measure of the strength of the forces. These constants are evaluated from measurements of the probability of an interaction when two particles meet (a term often used by physicists in this regard is the 'cross-section' of the interaction. The cross-section is measured in units of area, and the greater its value, the greater the probability of an interaction.) The coupling constant may also be derived from the average lifetime of a particle which decays spontaneously as a result of a certain interaction. Despite their name, 'coupling constants' are not all that constant. If we examine, for example, a number of weak interactions, we find that the coupling constants vary somewhat from one interaction to another. But on average, the ratios between the strength of the strong, electromagnetic, weak and gravitational forces were found to be: $1:10^{-2}:10^{-13}:10^{-39}$. In other words, the strong force is about 100 times stronger than the electromagnetic force, 10^{13} times stronger than the weak force, and 10^{39} times stronger than the force of gravity.

We will explain the above ratios with the help of an example. A neutron decays under the influence of the weak interaction to give a proton, an electron and a neutrino. The average lifetime of the neutron is 17 minutes. The 'lifetime' of the reverse interaction ($\bar{\nu}_e + p \rightarrow n + e^+$) is also 17 minutes; in other words, for the reverse interaction to take place, the neutrino must be in contact with protons for 17 minutes on average. Since the cross-section of the nucleus is about 10^{-8} of the cross-section of the atom, when the neutrino passes through matter it comes into contact with protons only 10^{-8} of the time. Therefore, for the interaction to take place, the neutrino must travel through matter for 17×10^{8} minutes on average. During this period the neutrino (which travels at the speed of light) covers a distance of 3×10^{16} kilometres. This is, then, the average thickness of the wall needed to stop a neutrino! Let us compare this with the photon absorption when light strikes an opaque substance (such as metal). Experiments show that the photon is absorbed by an atom after it has penetrated the substance to an average depth of 3×10^{-7} centimetres (about 10 atoms deep). The photon traverses this distance in 10^{-17} seconds. We must, however, take into account that the wavelength of a photon of visible light is about 5×10^{-5} centimetres, or about 1000 times the diameter of an atom (characteristic wavelengths of the neutrino are about one-thousandth of the diameter of an atom). The photon thus comes into simultaneous contact with $1000 \times 1000 \times 1000$ atoms while penetrating matter. If it were to come into contact with only one atom at a time, its absorption would take 10^9 times as long, or 10^{-8} seconds ($10^{-17} \times 10^9$). According to our calculations, this is the time interval necessary for the photon and atom to be in contact so that the photon will be absorbed (on condition that the atom *can* absorb a photon of such energy. The atoms in glass, for example, are not able to absorb photons of visible light and glass is thus transparent to visible light.) Many experiments have proven that this is also the time interval necessary for an atom to emit a photon of light (after the time it receives the necessary energy).

Now let's compare the results of both calculations. The neutrino must be in contact with a proton for an average of 17 minutes (about 1000 seconds) for the former to be absorbed due to the weak interaction. Only 10^{-8} seconds are required for a photon and an atom to be in contact to enable the former to be absorbed due to electromagnetic force. The absorption of the photon is thus 10^{11} times faster, just as expected according to the relative strengths of interactions mentioned above ($10^{-2}/10^{-13} = 10^{11}$). Similarly, we can compare the absorption of the pion by the nucleus, which takes place after a small number of collisions (due to the strong force), with the absorption of a muon by the nucleus, which is a weak interaction and a much rarer event (so rare that it does not prevent the muon from traversing the entire atmosphere). We should note that the spontaneous decay of the pion is a weak interaction, since the pion is the *lightest particle* which is affected by the strong interaction, and thus is unable to decay under its influence into lighter particles.*

* According to our modern view of the weak interactions, which stems from Weinberg–Salam electroweak theory, the reaction $\bar{\nu}_e + p \rightarrow n + e^+$ is actually a two-step process. First the anti-neutrino disintegrates into a virtual pair of a positron and a W^- ($\bar{\nu}_e \rightarrow e^+ + W^-$), then the W^- interacts with the proton ($W^- + p \rightarrow n$). The coupling constant of each of these processes is roughly 5×10^{-7}, hence the overall 'strength' of 10^{-13}.

See comment in text (end of section 6.2) concerning the carriers of the strong force.

Table 6.1. *The four basic interactions*

Type of interaction	Relative strength	Range (centimetres)	Carrying particle: Particle	Rest mass	Spin
Strong	1	$\sim 10^{-13}$	Mesons	$>10^2$ MeV	0
Electromagnetic	10^{-2}	Infinite	Photon	0	1
Weak	10^{-13}	$\sim 10^{-15}$	$Z^0, W^{\pm}$	$\sim 10^5$ MeV	1
Gravitational	10^{-39}	Infinite	Graviton	0	2

The coupling constant for gravity was calculated from Newton's equation. Its infinitesimal proportions, as compared to the coupling constants of the other forces, emphasizes the negligible importance of gravitation in particle physics.

Not all weak processes are as slow as the decay of the neutron. We have seen that the lifetime of a muon is 10^{-6} seconds, that of charged pions, 10^{-8} seconds and that of strange particles, about 10^{-10} seconds. The decay of all these particles is caused by the weak interaction (and we assume that in every case, the moment a particle is created, it is destined to decay). The great differences in the lifetimes are due to the fact that the rate of decay is influenced not only by the coupling constant, but by other factors as well, one of which is the difference between the mass of the decaying particle and that of its products; the mass of the neutron is only slightly greater than the mass of its decay products (i.e. the proton and the electron), and its decay is therefore particularly slow.

Characteristic lifetimes for particles which decay by electromagnetic interactions lie between 10^{-16} seconds ($\pi^0 \rightarrow 2\gamma$) and 10^{-19} or 10^{-20} seconds ($\Sigma^0 \rightarrow \Lambda^0\gamma$, $\eta^0 \rightarrow 2\gamma$). Decays caused by the strong force are faster by about three orders of magnitude. The lifetime of a particle enables us to determine with almost absolute certainty which interaction is responsible for its decay. In what follows, we will better understand just how the particle 'chooses' its manner of decay. The important properties of the four basic interactions are summarized in Table 6.1. It should be mentioned that the mesons, according to the modern theory, can be regarded as effective pseudocarriers of the strong force, at the nucleon level. The basic carriers of this force are the massless 'gluons' described in section 9.8.

6.3 The classification of particles

When the Russian chemist Dmitri Mendeleev first proposed his periodic table of elements in 1872, he could not have foreseen the revolution in physics and chemistry that would result from his modest discovery, that chemical elements 'fit' nicely into a table arranged according to rising atomic weight (or atomic number, in modern terms). Many scientists before him sought to arrange and classify the chemical elements in various manners, according to their density, colour, or physical properties. Mendeleev succeeded because he chose the correct way in which to arrange them, a way which later threw light on the internal structure of the atom itself.

The same is true with regards to elementary particles. Several systems of classification were proposed until we arrived at the system in use today, which best clarifies the similarities and differences between the various particles. This system is based on two properties of the particles: their spin and their susceptibility to the strong force.

We have already mentioned (in section 2.6) that *bosons*, those particles that have integer values of spin (0, 1, 2, etc. in $h/2\pi$ units) differ in their properties from *fermions* that have 'half-integer' spins ($\frac{1}{2}$, $\frac{3}{2}$, $\frac{5}{2}$, etc.). The main difference between these two groups is associated with Pauli's exclusion principle, which asserts that two *identical* fermions can never occupy the same vicinity. The electrons and the nucleons which build the atom are fermions (of spin $\frac{1}{2}$). The short-lived hyperons are also fermions; their similarity to the nucleons lies in the fact that we can ascribe a definite diameter (of the order of 10^{-13} centimetres) to each of these particles. However, even the electrons (which can be considered as points of mass) and even the massless neutrinos cannot pass through one another or even approach one another without limit. The Pauli principle dictates that each of these particles must occupy a defined region in space.

Pauli's principle, however, does not apply to the bosons (e.g. the photon which has spin 1, and the pions which have spin 0), and they may pass right through one another. Problems of over-crowding do not bother them at all – any number of bosons may be concentrated in a given region in space.

There is another difference between fermions and bosons. In the next chapter we shall see that fermions appear to be subjected to certain conservation laws that do not allow a fermion to be created or disappear without the simultaneous creation or disappearance of another fermion (for example, its anti-particle). The bosons, on the other hand, can be created and annihilated with no restriction (as long as the 'classic' conservation laws of energy, momentum and charge are not violated). These differences between the fermions and bosons create 'an apportionment of roles' between them: the particles that build the matter in our world are fermions, while the particles carrying the basic forces are all bosons.

The distinction between fermions and bosons divides all the particles into two groups. This, however, is not enough. A more successful division is obtained if we apply a further criterion of classification – the sensitivity of particles to the strong force. These two criteria put all the experimentally observed particles into four distinct groups: leptons, intermediate bosons, mesons and baryons.

Leptons are fermions which are unaffected by the strong force. Up until the mid-1970s, only four particles of this kind were known: the electron, the muon and two neutrinos (ν_e, ν_μ), as well as their four corresponding anti-particles. Today we know another pair of leptons, which will be described in Chapter 10.

The intermediate bosons are bosons which are unaffected by the strong force. They include the photon, the $W^\pm$ and Z^0 particles which carry the weak force and the graviton.

Mesons are bosons which are affected by the strong force. Of this group, we have thus far encountered the three pions, four kaons, and the eta meson (η^0). All have zero spin. Later on we will make the acquaintance of many additional mesons, of spin 1, 2, etc . . .

Baryons are fermions which are affected by the strong force. This group includes the nucleons, the hyperons and their respective anti-particles, as well as other, short-lived particles which you will encounter during the continuation of your reading.

The baryons and the mesons – the two groups of particles influenced by the strong force – are collectively known as the hadrons (*àdros*, in Greek, means strong). The names of the other groups are also derived from Greek: baryons meaning the heavy ones, mesons, the intermediate, and leptons, the light ones. At the time when the groups were named, the heaviest particles were among the baryons, and the lightest ones were leptons. However, many short-lived mesons which have later been discovered, as well as the additional lepton mentioned above, are heavier than the nucleons. Today the mass of a particle is not necessarily an indication of the group to which it belongs.

Table 6.2. *The classification of particles into four groups*

	Particles affected by the strong force (hadrons)	Particles unaffected by the strong force
Fermions (half-integer spin)	Baryons (p, n, Λ, Σ, Ξ, Ω, . . .)	Leptons (e, μ, ν, . . .)
Bosons (integer spin)	Mesons (π, K, η, . . .)	Photon (γ), graviton, W^+, W^-, Z^0

It should be emphasized that the various groups are not closed! Baryons and mesons decay in many cases to leptons. A photon of sufficient energy may turn into an electron–positron pair. In practically every interaction we have looked at so far, particles of more than one group have been involved. Nevertheless, from the present classification system several conservation laws were derived, as we shall see in the next chapter. Table 6.2 summarizes the definitions of the four groups of particles.

Chapter Seven

Conservation laws

7.1 Conservation laws and symmetries

What are the laws which rule our particle jungle? Is it possible to comprehend why one particle or another decays in a given manner, or to predict the results of a collision between two particles? The developers of quantum mechanics showed us the basic path which we must take: wave functions, wave equations, and so on. But the theory is incomplete, as we have no a-priori knowledge of the interparticle forces. Moreover, even when these forces become known, we often find that the calculations, in practice, are too complicated.

A number of simple laws, known as 'conservation laws', provide us with great assistance in both these concerns. Each law of this kind states that a certain physical quantity cannot change during the course of reactions between particles. The quantity remaining after the reaction must be identical to the quantity present beforehand. We can illustrate this with an example from the field of economics. In a country in which money is printed only at the rate at which it is taken out of circulation the general amount of money in circulation remains fixed. An industrial firm whose incomes exactly equal its expenditures maintains constant internal capital.

The physicists who dealt with classical physics (the physical theories prior to the appearance of relativity and quantum mechanics) were aware of a number of conservation laws, derived from decades of experimentation. Some of them, the law of charge conservation, for example, sprang up almost automatically. From the moment that electric charge was defined, it was patently obvious that it cannot be created, nor does it disappear; in other words, its quantity in the universe is conserved. The birth pains of other conservation laws were more prolonged, and each was actually based on a proper definition of the conserved quantity. The law of conservation of energy, for example, weathered many changes, since it is based on a proper definition of all forms of energy, some of which – such as heat or mass – are not clearly related to mechanical energy.

In classical physics, conservation principles are used to simplify difficult calculations. If one is to deal with a complicated multi-step process, it is helpful to know that certain quantities remain constant throughout the whole process. The relative importance of conservation laws in classical physics was, at any rate, secondary. Basic laws of physics such as Newton's laws in mechanics and Maxwell's electromagnetic equations deal in dynamic or changing, rather than constant, quantities.

In modern physics, not only were several conservation laws added, but their relative importance increased as well. And this is no surprise: in a world ruled by probability, where laws such as those of Newton are unable to predict just what will happen, laws which determine *what cannot happen* become increasingly important. If we know that no process which contradicts any conservation law can take place, we may infer that for any process which *does not* contradict any conservation law, a certain probability exists that it will occur. The conservation laws are among the most basic laws of nature. Knowing all the conservation laws ruling a certain process is tantamount to being able to determine the *probability* of any particular result of the reaction or scattering. This is the quantum version of kinematics. Moreover, some conserved quantities, such as electric charge, have a dynamical role: they are the source of a force. Identifying the conservation laws is therefore the clue to dynamics.

The statement that whatever does not contradict the conservation laws does actually take place has great significance. Let us assume that a given particle can decay into other particles without breaking a single conservation law. Will it actually do so? The answer is yes! All of the experimental data at hand indicate that it will do so as soon as possible! This is the reason why most particles are not stable; the few particles which are stable are those which *cannot* decay, because their decay will break one conservation law or another. We will encounter some enlightening examples as we progress.

Let us now summarize this important change in outlook: instead of the set of deterministic laws predicting exactly what will occur, which ruled in classical physics, modern physics is ruled by conservation laws, which determine what *cannot occur*. And according to the present approach, any process which does not contradict these laws does indeed take place.

Certain quantities remain constant throughout all interactions while others remain constant in some interactions but may change in others. Before we proceed to list them, let us have a look at the interesting relationship between conservation laws and symmetries.

Have you ever asked yourself, 'What is the meaning of the word symmetry?' Fig. 7.1 presents several symmetric forms. When we say that a certain pattern is symmetric, we mean that its shape will not change if we turn it upside down, reflect it in the mirror, and so on. In each case, we can define those operations with respect to which the form is symmetric. For example, Fig. 7.1*c* is symmetric with respect to a 180° rotation around its centre. Physicists have extended the concept of symmetry beyond its geometrical meaning. One can say that the algebraic equation $a + b = c + d$ is symmetric with respect to the operation:

$$a \rightarrow -c \quad c \rightarrow -a$$

In other words, if we replace every 'a' with '$-c$', and every 'c' with '$-a$', the equation will not change. Another concept which is frequently used in this regard is *invariance*. We say that the equation is invariant (does not change) with respect to the operation mentioned above. It may be shown that almost every conservation law stems from a certain elementary symmetry of nature, or, in other words, from a certain principle of invariance; from the fact that the laws of nature do not change under a certain imaginary operation.

These relations between conservation laws, symmetries and invariance principles were pointed out by Emmy (Amalie) Noether. She formulated a theorem stating that from the fact that the laws of physics are not affected by some operation or 'displacement', a conservation law for some physical quantity originates.

The proofs of these relations are far from being simple – they are carried out using advanced mathematical methods, and we will not go into detail here. We will, however, indicate the symmetry principles which lie behind some conservation laws.

The conservation laws which were known in classical physics are:

(1) conservation of energy and mass;
(2) conservation of linear momentum;
(3) conservation of angular momentum;
(4) conservation of electric charge.

We will now review these conservation laws, one by one.

Among the first to recognize the importance of symmetry considerations and invariance principles in physics was Karl Gustav Jacobi, one of the outstanding personalities in mathematics and physics in the nineteenth century. There is practically no field of mathematics or mechanics on which Jacobi did not leave his stamp.

Simultaneously with Jacobi, the concept of symmetry in physics was developed by the great Irish mathematician William Rowan Hamilton, and later on by the French physicist Pierre Curie (husband of Marie Curie). At the beginning of the twentieth century the idea was perfected by the German mathematician David Hilbert and his student Amalie (Emmy) Noether, an outstanding mathematician, daughter and sister of mathematicians. Hilbert wanted to arrange a professorship for Emmy Noether at Göttingen, which was then the foremost centre for mathematics in Germany, but because she was a woman there was strong opposition to such an appointment. Hilbert managed to overcome all the official obstacles, but there remained one insuperable problem: the senate building of the university had only one lavatory – for men – so how could a woman serve on the senate? In 1933 Emmy was forced to flee to the USA, where she died two years later.

(*a*)

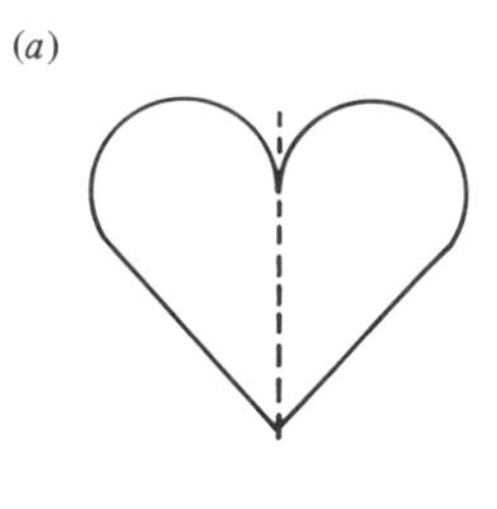

(*b*)

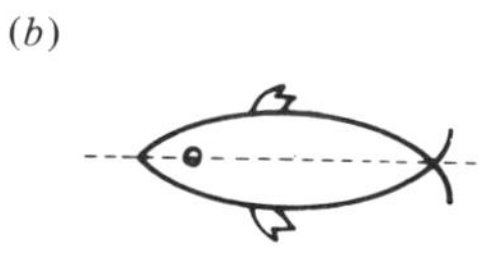

(*c*)

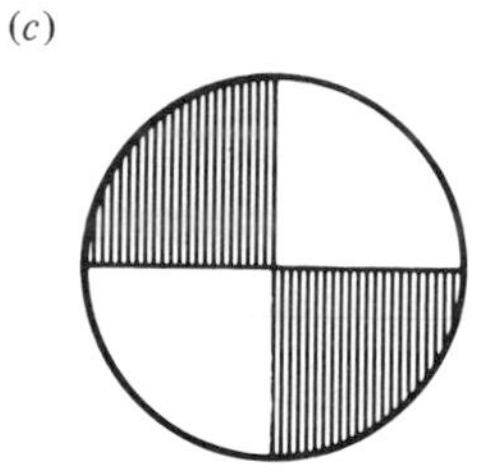

(*d*)

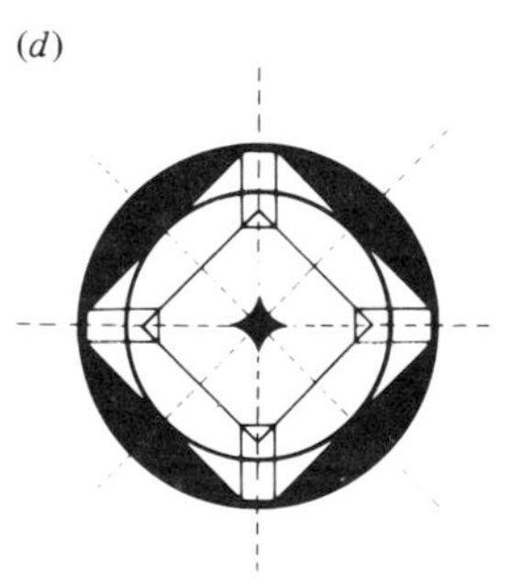

Figure 7.1. Symmetric shapes. Shapes (*a*) and (*b*) are symmetric with respect to reflection with respect to the dashed line, which is called 'the symmetry axis'. Shape (*c*) is symmetric with respect to a 180° rotation around the centre, and (*d*) is symmetric with respect to rotations of 90°, 180° and 270° around the centre, as well as to reflection with respect to each of the dashed lines.

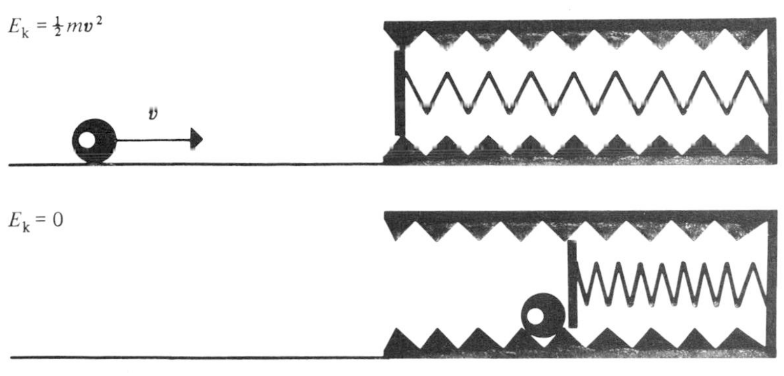

Figure 7.2. The kinetic energy, E_k, of the moving body is converted into the potential energy of the compressed spring.

7.2 The conservation of energy and mass

The law of energy conservation is considered to be one of the most important laws of nature. Generations of physicists were involved in its formulation. Since energy can assume a variety of shapes, this conservation law was altered whenever a new type of energy was discovered. Its simplest form – the conservation of mechanical energy – was already known in the seventeenth century, when it was found that under certain conditions the sum of the kinetic and potential energies of a body remains constant (see Fig. 7.2). Any moving body possesses kinetic energy which is given (at velocities much smaller than that of light) by $\frac{1}{2}mv^2$, where m is the mass and v the velocity of the body. The potential energy of a body depends on the external forces acting on it. For example, a body poised h metres above the ground possesses a potential energy of mgh with respect to the ground, due to the force of gravity, where g is the free-fall acceleration (see Fig. 7.3). A modern example for the conservation of mechanical energy is a proton or an electron which enters a one-stage linear accelerator with a negligible kinetic energy and a potential energy equal to eV (where V is the accelerating potential difference) and goes out at the other end with a substantial kinetic energy. In the non-relativistic case the following equation holds: $\frac{1}{2}mv^2 = eV$ and thus the final velocity of the particle is given by $v = (2eV/m)^{1/2}$.

In most everyday systems, mechanical energy is not conserved, owing to frictional forces. However, after heat was recognized as a form of energy, a more general conservation law was formulated, which assures us that in this case mechanical energy is merely

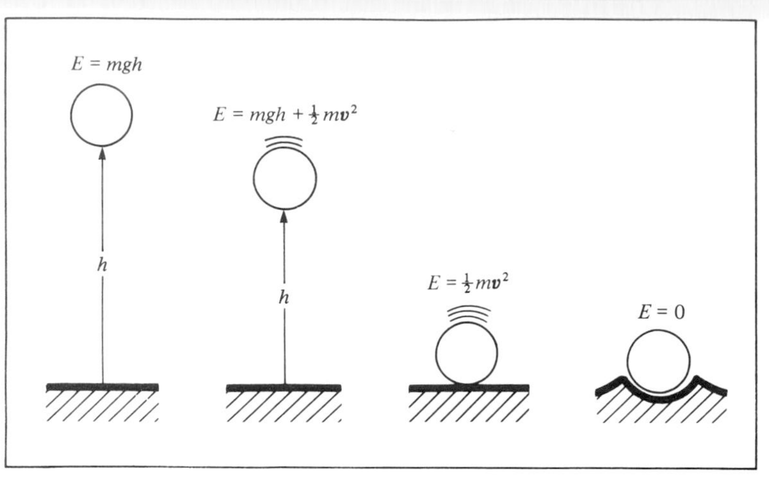

Figure 7.3. A mass m at a height h above the ground has a potential energy of mgh with respect to the ground. All the energy of a falling body is potential energy at the beginning of the fall, and kinetic energy when it is near the ground. When the body hits the ground, its energy is converted into heat.

transformed to heat (or thermal energy) which causes the temperature of the system to rise. In the last century it was found that thermal energy is none other than the kinetic energy of the molecules within the substance. Thus, when dealing with molecules, atoms or sub-atomic particles, one can usually refer to their energy as solely mechanical energy, with no need to speak of heat.

A dramatic extension of the law of energy conservation was presented in the twentieth century when it was found that mass is a form of energy and that in nuclear processes and in reactions between particles mass may be converted into energy and energy into mass (see section 2.1). One general conservation law – that of mass and energy – has thus replaced the two separate conservation laws. According to this law, the sum of energy and mass in a *closed system* (when the mass is expressed in units of energy, according to Einstein's equation $E = mc^2$) is constant and does not change with time. We emphasized the words 'closed system'. One can, of course, increase the mass–energy content of a system by putting extra mass in, or by exerting external force on its components and thus increasing its energy (resulting in the decrease of the mass–energy content of another system).

We have mentioned at the beginning of the chapter the basic link between conservation laws and symmetry principles. The symmetry principle from which the energy and mass conservation law stems states that physical laws are invariant with respect to time. In other

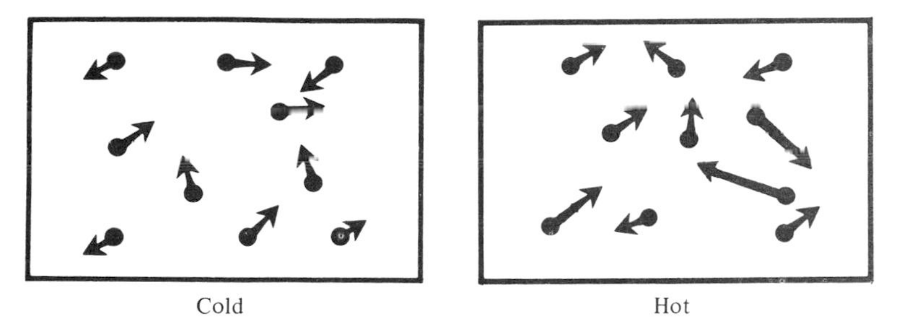

Figure 7.4. Heat is the kinetic energy of atoms and molecules. The average kinetic energy (and average velocity) of the molecules is higher in a hot gas than in a cold gas. The arrows in the figure represent the velocity vectors of the gas molecules at two different temperatures.

words, any physical law which applies today will apply tomorrow as well as 1000 years hence. It can be shown that from this simple principle the conservation law can be deduced, and that deviations from this principle would allow the creation of mass or energy out of nothing.

7.3 The conservation of linear momentum

The equations in the text give only the magnitude of the momentum, which, being a vector, has also a direction in space.

The linear momentum ($\boldsymbol{p}$) (or simply – the momentum) of a mass moving at a velocity which is low compared with that of light is defined as the product of mass and velocity: $\boldsymbol{p} = m\boldsymbol{v}$. This equation is valid at relativistic velocities as well (i.e. velocities approaching c – the speed of light), but then m is not the rest mass of the particle, but rather its relativistic mass: $m(\boldsymbol{v}) = m_0/(1 - v^2/c^2)^{1/2}$. The momentum of a massless particle, such as the photon, is defined as $\boldsymbol{p} = E/c$, E representing the energy of the particle. A more general definition is: $\boldsymbol{p} = (E^2/c^2 - m_0{}^2c^2)^{1/2}$ where E represents the total relativistic energy of the particle (kinetic energy and energy resulting from mass), and m_0 is the rest mass. This equation holds for massless particles ($m_0 = 0$) as well as for particles bearing mass, regardless of their velocity. In fact, when the velocity of the particle is low, this formula simplifies to $\boldsymbol{p} \approx m_0\boldsymbol{v}$.

One should remember that the momentum is a vector quantity, meaning that its direction is just as important as its magnitude (energy, on the other hand, has no direction and is thus classified as a *scalar* quantity). The direction of the momentum of a single particle is the direction of its motion. To calculate the momentum of a group of particles, we must add the vectors representing their individual momenta in a certain geometric manner (see Fig. 7.5). According to the law of momentum conservation, the total linear momentum of a system is conserved, as long as no external forces

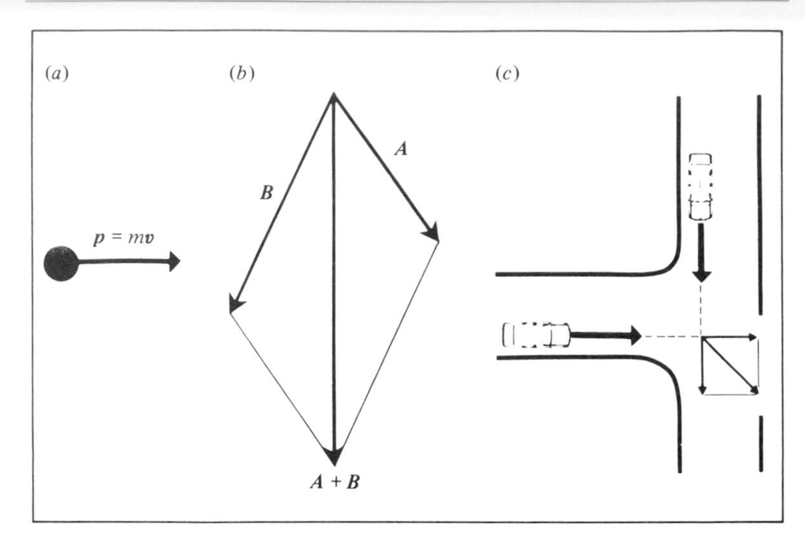

Figure 7.5. (a) The momentum is a vector; i.e. it has both magnitude and direction. The momentum of a particle of mass m and velocity $\boldsymbol{v}$, is represented by an arrow pointing in the direction of the velocity, whose magnitude is $m\boldsymbol{v}$. (b) Adding up vectors by the parallelogram method. The sum of two vectors is the diagonal in the parallelogram, the sides of which are those vectors. (c) In order to find the sum of momenta of two cars approaching a crossroad, one has to bring the two momentum vectors to a single point. The diagonal of the parallelogram they form is the resultant momentum. This diagonal represents the direction of motion if the cars collide and stick together.

are acting on the system. It should be stressed that when external forces are acting on a system, in addition to the internal forces between its various components, then its momentum is not conserved. When a body falls owing to the force of gravity, for example, its momentum grows continuously.

The recoil of a cannon after firing a shell is one example of the conservation of linear momentum. Before firing the shell, the momenta of both cannon and shell are zero. Since the fired shell acquires momentum in a given direction, the cannon must acquire an equal momentum in the opposite direction so that their sum remains zero. In particle physics, the conservation of momentum dictates that the total momentum of a group of particles prior to a reaction must be equal in magnitude and direction to their total momentum following the reaction (providing that no *external* forces act on the system).

The use of the momentum conservation law simplifies the analysis of reactions between particles. For instance, when a particle at rest splits into two particles they will move in opposite directions (Fig. 7.6a). The ratio of their speeds will equal the inverse of the ratio of

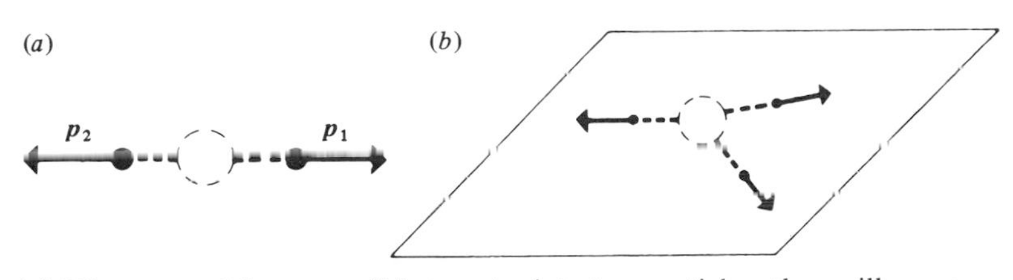

Figure 7.6. (*a*) When a particle at rest disintegrates into two particles, they will move in opposite directions. The lighter particle will move faster. (*b*) When a resting particle disintegrates into three particles, they are constrained to move in the same plane.

their masses (the lighter particle will move faster). When a stationary particle splits into three particles, they must, due to momentum conservation, move in the same plane (Fig. 7.6*b*. This principle has quite general applications since any particle may be regarded as a stationary particle, provided that the calculations are made in a coordinate system which moves together with the particle.) But the most important and most basic law derived from momentum conservation is that no particle can spontaneously decay into particles whose mass exceeds its own. At first glance, it would appear possible for some of the kinetic energy of a rapidly moving particle to be converted into mass, thus enabling a spontaneous decay into particles of greater mass. This, however, is not possible since it can be shown that either the law of conservation of linear momentum or the law of conservation of energy would be violated in such a case. For this reason, a photon cannot turn into a positron–electron pair, except in the immediate vicinity of a nucleus which can absorb the excess momentum, as we have seen in section 3.2.

When a rapidly moving particle collides with another particle, heavier particles can be produced (in contrast with a spontaneous decay), but even then only part of the kinetic energy can be transformed into mass or the conservation of momentum would be violated. For example, when a fast-moving proton collides with a stationary proton, another pair of nucleons may be produced:

$$\mathrm{p} + \mathrm{p} \rightarrow \mathrm{p} + \mathrm{p} + \mathrm{p} + \bar{\mathrm{p}}$$

This is how anti-protons are generated in accelerators. The new proton–anti-proton pair is created from the kinetic energy of the moving proton. The mass of this pair is less than 2 GeV. Even so, in order to create the pair, the colliding proton must be accelerated until it attains a kinetic energy of about 6 GeV, and only one-third of this kinetic energy would actually be converted to mass. We should point out here that if the two original protons moved towards one another with equal velocities, all their kinetic energy could be converted into mass. This principle is exploited in a storage ring or collider accelerator (see section 8.4).

Another result of the conservation of momentum is that a particle cannot decay into a *single* particle lighter than itself. For example, the spontaneous process $\Lambda^0 \not\to n$ is impossible, since the difference in mass would have to be converted into kinetic energy of the neutron, and calculations show that the momentum balance would be upset.

The symmetry principle from which the law of momentum conservation stems states that the laws of nature are invariant with respect to translations in space. A physical law which is true for a certain point in space holds for another point as well. If we develop a physical equation for a given coordinate system, its form will not change, in principle, if the coordinate system is shifted.

It is quite surprising to learn that such basic laws of physics as those of momentum and energy conservation are based on simple truths which appear almost self-evident. One stems from the homogeneity of space, i.e. from the fact that space is uniform throughout, and the second from the homogeneity of time, i.e. from the fact that the laws of physics are not dependent on the date. This discovery fulfils one of the principal aims of physics – to base the important laws on principles which are as simple and as general as possible.

7.4 The conservation of angular momentum

Angular momentum plays the same role in rotational motion as linear momentum plays in motion along a straight line. The angular momentum of a particle orbiting in a circle is a vector, $\boldsymbol{L}$, the magnitude of which is the product of the mass m, the velocity $\boldsymbol{v}$ and the radius of the circle r: $\boldsymbol{L} = m\boldsymbol{v}r$. The direction of the vector is perpendicular to the plane of the rotation as shown in Fig. 7.7. The

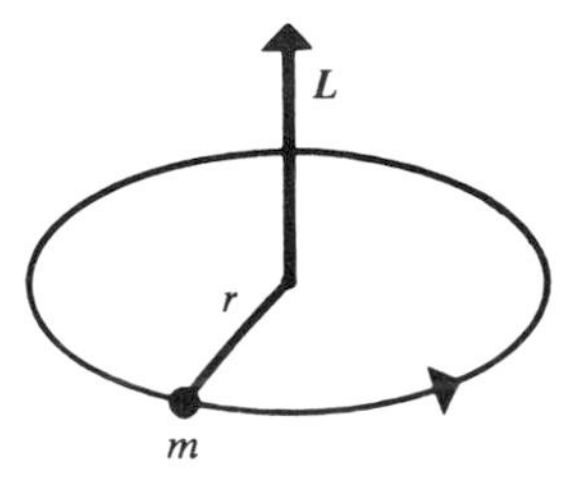

Figure 7.7. The angular momentum of a particle of mass m, which moves in a circle of radius r, at a velocity $\boldsymbol{v}$, is a vector $\boldsymbol{L}$, the magnitude of which is: $\boldsymbol{L} = m\boldsymbol{v}r$. $\boldsymbol{L}$ is perpendicular to the plane of rotation, and is related to the trend of rotation, as the direction of advance of a right-handed screw is to the direction of its revolution. The magnitude of $\boldsymbol{L}$ can also be written as: $\boldsymbol{L} = 2\pi f m r^2$ where f is the frequency of rotation.

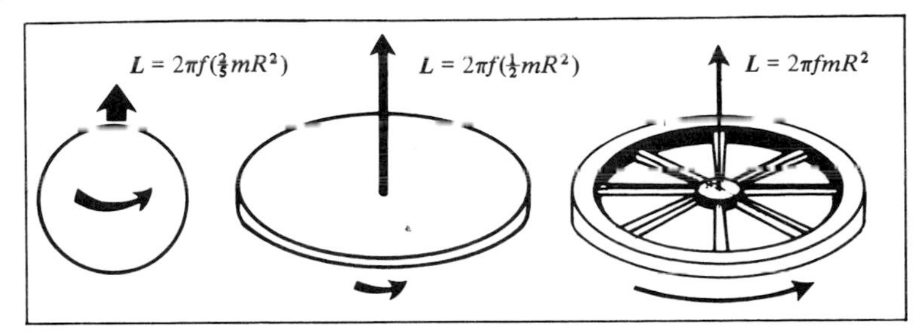

Figure 7.8. The angular momenta of three simple bodies of mass m and radius R; a uniform sphere, a disc and a wheel.

angular momentum of a system of particles is the vector sum of the angular momenta of the various particles. The angular momentum of a rotating rigid body can be calculated by treating the body as a large collection of grains, each with a tiny mass and moving in a circular orbit. The resulting equations for three simple bodies are presented in Fig. 7.8.

The importance of angular momentum lies in the fact that, similarly to linear momentum, this quantity is strictly conserved when no external forces are acting on the system.

This conservation law explains a few interesting physical phenomena. For example, at the turn of the seventeenth century, Johannes Kepler discovered that the planets are orbiting the sun in elliptical paths and that each of them speeds up as it approaches the sun and slows down as it moves away from it (see Fig. 7.9). This phenomenon is easily accounted for by the law of conservation of angular momentum: the angular momentum of the planet is proportional to its velocity and its distance from the sun. The angular momentum must remain constant; therefore, when the distance decreases the velocity must increase, and vice versa.

Here is another example from everyday life: you can ride your bicycle without falling partly because each spinning wheel has angular momentum the direction of which is perpendicular to the

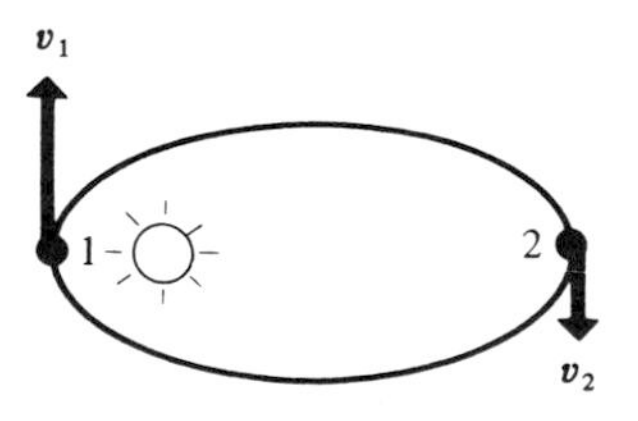

Figure 7.9. The planets circle the sun in elliptical orbits. At point 1 the distance, r, between the planet and the sun is small and the velocity, v, of the planet is high. At point 2, the distance is large while the velocity is low. The angular momentum, whose value is mvr, remains constant.

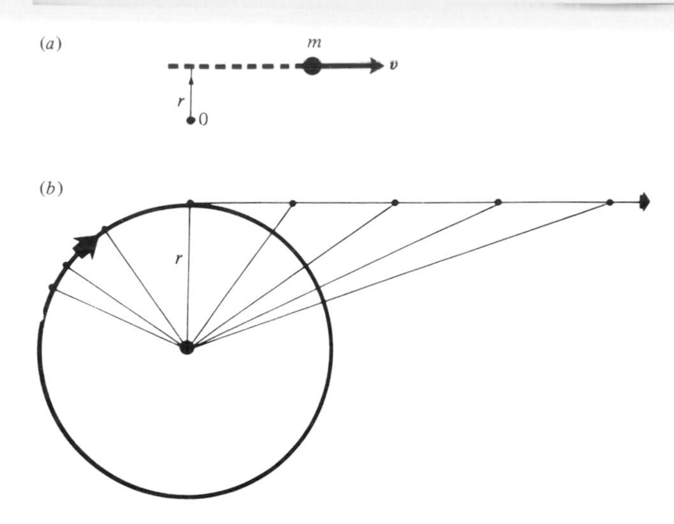

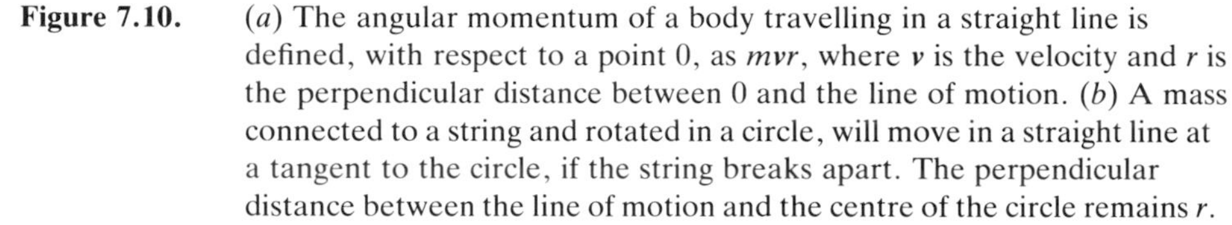

Figure 7.10. (*a*) The angular momentum of a body travelling in a straight line is defined, with respect to a point 0, as mvr, where v is the velocity and r is the perpendicular distance between 0 and the line of motion. (*b*) A mass connected to a string and rotated in a circle, will move in a straight line at a tangent to the circle, if the string breaks apart. The perpendicular distance between the line of motion and the centre of the circle remains r.

plane of the wheel. Owing to the conservation law, the direction of this vector is conserved, and this contributes to the stability of the bicycle.

Another fine example of this conservation law is the revolving ice skater: when he spreads his arms during a fast spin, he slows down immediately; and when he folds his arms back across his chest, he once again revolves much faster. The change in velocity compensates for the fact that a part of the mass changes its distance from the axis of rotation, in such a way as to keep the angular momentum unchanged.

It should be mentioned that although it is of special importance in rotational motion, angular momentum can be defined for a body travelling in a straight line as well, as shown in Fig. 7.10.

Angular momentum in quantum mechanics

Angular momentum bears special importance in quantum mechanics since it is often convenient to characterize states of atomic systems such as atoms, molecules or nuclei by their energy and angular momentum. The angular momentum (like the electric charge) has a basic natural unit. This unit is Planck's constant divided by 2π, and it is frequently denoted $\hbar$.

$$\hbar = h/2\pi = 1.0545 \times 10^{-34} \text{ joules} \cdot \text{second}$$

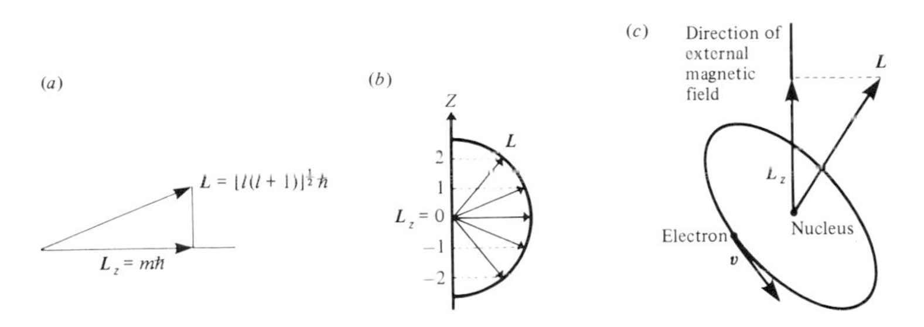

Figure 7.11. (*a*) The projection of the angular momentum vector, $\boldsymbol{L}$, in a defined direction (such as the direction of a magnetic field) is indicated by $\boldsymbol{L}_z$. This projection is always a multiple of $h/2\pi$: $\boldsymbol{L}_z = mh/2\pi$, where m takes integer values between $+l$ and $-l$ in steps of 1. The total number of allowed values of $\boldsymbol{L}_z$ is $2l + 1$. (*b*) If, for example, $l = 2$, then $\boldsymbol{L} = (h/2\pi)\sqrt{6}$, and $\boldsymbol{L}_z$ can take the values 2, 1, 0, −1, −2 (in the 'natural' units $h/2\pi$). (*c*) The vector $\boldsymbol{L}$ and its projection $\boldsymbol{L}_z$ in Bohr's model of the hydrogen atom. In this model, the angular momentum of the electron is associated in a simple way with its rotation around the nucleus.

Angular momentum differs in this respect from energy and linear momentum which do not have basic natural units. We have mentioned (Chapter 2) that one of the basic assumptions of Bohr's model was that the angular momentum of the electron in the hydrogen atom should always be an integral multiple of $h/2\pi$: $\boldsymbol{L} = nh/2\pi$ (n = 1, 2, 3, . . .). The Schrödinger equation gives a somewhat different result: the angular momentum associated with the motion of an electron around the nucleus (orbital angular momentum) is quantized, i.e. it can take only certain discrete values, but those values are given by a slightly more elaborate equation:

$$\boldsymbol{L} = (h/2\pi)(l(l+1))^{1/2} \quad (l = 0, 1, 2, \ldots)$$

The number of allowed orientations of the angular momentum vector with respect to a defined direction is $2l + 1$.

Note that according to this equation, $\boldsymbol{L}$ *is not* an integral multiple of $h/2\pi$. Quantum mechanics tells us that this equation is valid for atoms, molecules and macroscopic bodies as well, but, as already mentioned, $h/2\pi$ is such a small number that only in atomic systems the quantization of angular momentum can manifest itself experimentally. Were $h/2\pi$ much bigger, we would find that a top cannot rotate at any speed, but only at certain discrete angular velocities.

Another notable consequence of the Schrödinger equation is that there exist certain limitations on the *direction* of the angular momentum. If a system is subjected to an electric or magnetic field, or to any other influence which defines some axis in space, then the angular momentum of the system can align itself only in such directions that its projection on this axis will be an integral multiple of $h/2\pi$ (see Fig. 7.11). It can be shown that in any measurement of

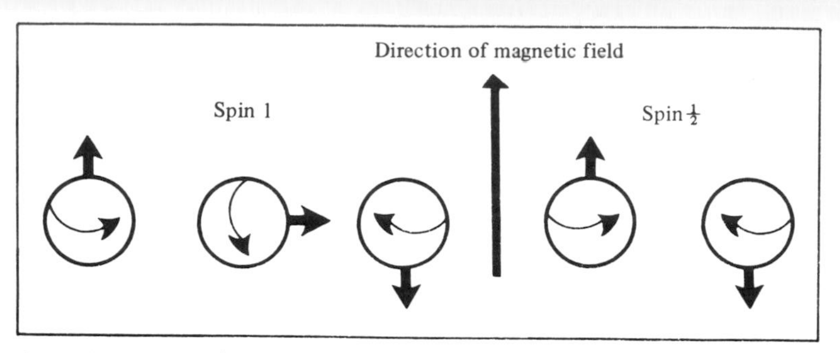

Figure 7.12. A particle of spin ½ in a magnetic field has two possible states: with the spin parallel or antiparallel to the direction of the field. A particle of spin 1 has three possible states. In this schematic description the spin is shown as arising from an actual rotation motion. In contrast to the more accurate picture of Fig. 7.11*b*, here the spin angular momentum is represented by a vector with the length of 0, ½, 1, etc. (in $h/2\pi$ units), which can be oriented in the exact direction of the field. This simplified approach is adequate for our needs.

angular momentum a preferred direction in space is determined and the measured quantity is actually the maximal projection of the angular momentum on this direction. Thus the *measured* value of the angular momentum projection is an integral multiple of $h/2\pi$, although $(l(l + 1))^{1/2}$ is not an integer.

A prominent difference between Bohr's model and quantum mechanics is that according to the latter the angular momentum of the electron in the hydrogen atom can be zero (if $l = 0$) which is impossible in Bohr's model. In fact, no classical model can ascribe zero angular momentum to an electron revolving around the nucleus. Experimental results have fully supported the prediction of quantum mechanics: in the lowest energy state of the hydrogen atom the angular momentum of the electron was found to be zero.

We already know that each particle has an intrinsic angular momentum or spin. The spin has all the properties of regular angular momentum except for the fact that its measured values can be 0, ½, 1, $\frac{3}{2}$, . . . (in natural units) instead of integer values. According to the quantization rule, a particle of spin ½ in a magnetic field has two possible states while a particle of spin 1 has three possible states (Fig. 7.12). The difference between one allowed projection and the next is always one unit of $h/2\pi$, both in the case of orbital angular momentum (Fig. 7.11) and in the case of spin.

The law of conservation of angular momentum applies to spins as well, and affects processes of creation and annihilation of particles. For example, when a photon turns into an electron–positron pair,

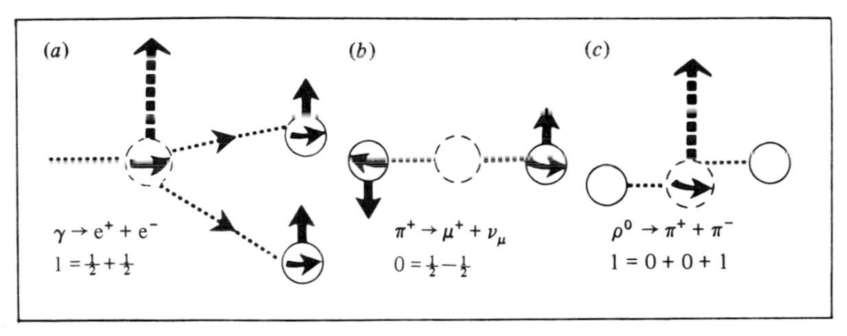

Figure 7.13. In the decay of a particle of (*a*) spin 1 or (*b*) spin 0 into two particles of spin $\frac{1}{2}$, the original spin angular momentum is conserved through the spins of the products. When a particle of spin 1 decays into two particles of spin zero (*c*) the original spin is converted into the orbital angular momentum of the relative motion between the decay products. (The ρ^0 is a short-lived meson which we shall meet later.)

the spins of the produced particles must be in the same direction as the spin of the original photon so that the total angular momentum after the interaction ($\frac{1}{2} + \frac{1}{2}$) will be the same as before it (Fig. 7.13*a*).

When a particle with zero spin decays into two particles, each with a spin of $\frac{1}{2}$, for example $\pi^+ \rightarrow \mu^+ + \nu_\mu$, the spins of the products are opposite in direction so that their overall spin remains zero (Fig. 7.13*b*). A particle of integral spin (1, 2, . . .) can decay into two particles, each of zero spin, which revolve around one another, or move on parallel straight lines. In this case, the angular momentum of the spin is converted into orbital angular momentum (Fig. 7.13*c*).

To conclude our discussion of angular momentum it should be mentioned that the conservation of angular momentum stems from the symmetry of space with respect to rotations. The form of an equation developed for a certain coordinate system will not change if we rotate the coordinate axis. It also means that there is no preferred direction in space. In empty space where no fields of force prevail, there is neither 'up' nor 'down' and the properties of a moving particle are in no way affected by the direction of its motion.

7.5 The conservation of electric charge

According to this conservation law, electric charge is neither created nor destroyed except in equal amounts of positive and negative charge. In an equation which describes a reaction between particles, the sum of charges on the left side of the equation must equal the sum on the right side, where in the calculation of the sum we add the positive charges and subtract the negative ones. For example, in the reaction:

$$p + p \rightarrow p + n + \pi^+ + \pi^+ + \pi^-$$

the sum of charges on each side of the equation is +2. In the reaction $\pi^- + p \rightarrow \Lambda^0 + K^0$, the sum of the charges on either side is zero. If an equation is not balanced with respect to charge, i.e. if the total charge on its right side is not equal to the total charge on its left side, this equation describes an impossible process.

Since any electric charge is an integral multiple of the basic unit charge, the total charge of a system can always be expressed as an integer – positive, negative or zero. The law of charge conservation is an example of conservation laws that are called counting rules or additive rules. We shall meet similar counting rules later. The stability of the electron results directly from the conservation of charge. All the particles which are lighter than the electron (the neutrinos and the photon) are not charged and thus the electron cannot decay spontaneously without breaking the law of conservation of charge.

What is the symmetry principle associated with the conservation of charge? It can be shown that this law stems from the fact that the equations of electromagnetism are not dependent on the definition of the zero point of electric potential. If we have a battery of 1.5 volts we can define the potential at the negative electrode as zero, and that at the positive electrode as +1.5 volts. On the other hand, we can also define the potential at the positive electrode as zero, and that at the negative electrode as −1.5 volts. And we can just as easily define the potentials at the electrodes as +0.75 and −0.75 volts respectively.

New conservation laws

We have reviewed the four conservation laws of classical physics: the conservation of charge, of energy and mass, of linear momentum and of angular momentum. These laws prevailed in particle physics as well. However, from the fact that there are many processes which do not contradict any of the classical conservation laws, and yet have never been observed, physicists came to the conclusion that interactions between particles obey *additional conservation laws* which are not revealed through the behaviour of macroscopic objects. These may be ad-hoc approximate rules, or they might be very fundamental. (It is believed that the more fundamental ones have dynamical roles, like the electric charge.) In any case, in order to formulate these new laws, novel 'charges' or quantum numbers had to be defined. These additional conservation laws and 'charges' will be described in the following sections.

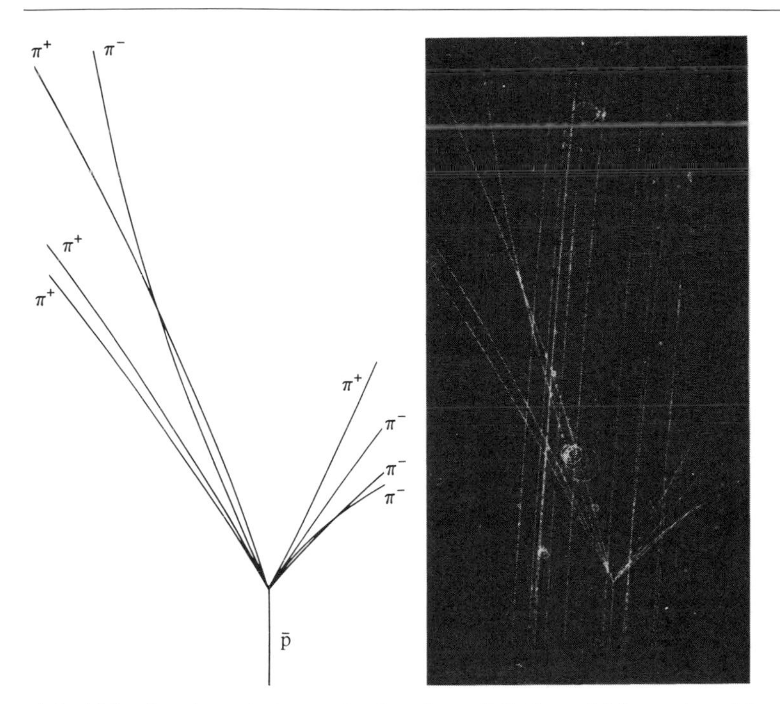

Figure 7.14. A bubble chamber photograph of a $p\bar{p}$ annihilation which produced four π^+, four π^-, and an unknown number of π^0. The charges of the pions were determined from the curvature of their tracks in the magnetic field which prevailed in the chamber. The parallel, almost straight, paths are those of anti-protons. (Courtesy Brookhaven National Laboratory, Upton, New York.)

7.6 Conservation of baryon and lepton numbers

Conservation of the baryon number

A multitude of particles may be produced in a collision between two protons, but mesons can never be the sole products. Reactions such as $p + p \not\to \pi^+ + \pi^+ + \pi^0$ have never been observed, even though they do not contradict any of the conservation laws discussed thus far. On the other hand, mesons can be the sole products of an interaction between a proton and an anti-proton (see Fig. 7.14). It appears as if another conservation law, similar to the law of conservation of charge, exists which prevents the first process from occurring, yet permits the second. An investigation of all the reactions in which baryons participate showed that they all obey a conservation law which we shall call: the conservation of baryon number. To each baryon (p, n, Σ, Λ, etc.) we ascribe the baryon number $A = +1$ and to each anti-baryon, the baryon number $A = -1$. The mesons (and the leptons as well) are given the baryon number 0. (The baryon number of a nucleus is its mass number. 'A' therefore denotes both the nuclear mass number and the baryon number.) The conservation law of the baryon number is a simple 'counting rule', similar to the conservation law of charge. It claims that in any interaction, strong, electromagnetic or weak, the sum of the baryon numbers of the products must equal that of the original particles. It is easy to see that this conservation law forbids the reaction $p + p \not\rightarrow \pi^+ + \pi^+ + \pi^0$ because the total baryon number on the left side of the equation is 2, and on the right side is 0. The reaction $p + \bar{p} \rightarrow \pi^+ + \pi^- + \pi^0$ on the other hand is feasible, because on both sides $A = 0$ (the baryon number of $\bar{p}$ is -1). Investigations of very many reactions in accelerators and cosmic rays have never discovered up to now a single process which violates the conservation of baryon number. For instance, the reaction $p + p \not\rightarrow p + p + n + n$, which is forbidden by this law, has never been detected even when the kinetic energy of the colliding protons was adequate for producing additional nucleons. On the other hand, the reactions $p + p \rightarrow p + p + n + \bar{n}$ and $p + p \rightarrow p + p + p + \bar{p}$, which do not violate the conservation of baryon number ($A = 2$ on both sides), occur frequently at adequate energies.

Hermann Weyl, in 1929, was the first to ask why protons in the atom do not disintegrate by absorbing the electrons. E. C. G. Stueckelberg in 1938 and Eugene P. Wigner in 1949 proposed that the proton is stable because of a conserved quantity which eventually was called *baryon number*.

Because the baryon number must be conserved, baryons are created and disappear in pairs. When a baryon comes into the world, an anti-baryon must be born along with it (though not necessarily its specific anti-particle). And when a baryon leaves the world without another baryon taking its place, it takes an anti-baryon along as well. The overall number of baryons in the world, less the number of anti-baryons, always remains constant.

The law of conservation of baryon number is responsible for the stability of the proton. Looking at Table 5.2 you will see that the decay products of each baryon include a lighter baryon, and since the proton is the lightest of the baryons, it cannot decay to other baryons. On the other hand, it cannot decay to mesons or leptons without violating the conservation of baryon number. Baryon number conservation, however, is an *ad hoc* law, and we do not know whether it is absolute or only approximate. Indeed some of the attempts to build a theory which unifies the strong force with the electroweak force (see section 6.1) lead to the startling prediction that the proton is not stable but would decay after a very long lifetime to lighter particles (e.g. $p \rightarrow e^+ + \pi^0$) by violating the conservation of baryon number. The simplest version of this theory, called 'minimal SU(5)', prescribes that the average lifespan of the proton should be about 2.5×10^{31} years or less.

The problem of proton stability can be examined experimentally by using a very large accumulation of protons (a large volume of water, for instance) surrounded by detectors which are sensitive to the expected products of the proton decay. By watching the pile of protons for an adequate period of time, one can verify or rebut the assumption of proton instability.

As Maurice Goldhaber put it 'we know in our bones' that the proton's lifetime is very long if not infinite. If the lifetime were shorter than 10^{16} years, the radioactivity stemming from the decay of protons in our body would imperil our lives.

Let's assume that the proton is indeed not stable and that half of the protons in a given sample will decay within a period of the order of 10^{31} years. If we have in our tank of water 10^{32} protons (some 300 tonnes of water) about *ten* protons should decay in the course of one year, and we would be able to detect their decay products. If there is a negative result, i.e. no proton decays have been detected, one can easily evaluate an experimental lower limit to the proton lifetime.

Planning such an experiment involves several practical problems which have to be solved. Since the event one tries to detect is so rare, the set-up must be protected from cosmic rays that may initiate other, less rare, events which might be mistaken for proton decays. An effective way to do this is to place the tank and detectors deep in the ground. The best shielding, however, is of no avail against neutrinos, and a method should be contrived to discern between events caused by neutrinos and true spontaneous proton decays.

Actually, an indication that the conservation of baryon number may be only an approximate law came from cosmology. If the universe was created out of nothing but gravitational energy (gravitons), it should contain matter and anti-matter in equal amounts. If there are no galaxies made of anti-matter (atoms made of anti-protons, anti-neutrons and positrons) as is now believed, the baryon number must have been violated in the early universe, to yield the excess of matter over anti-matter that led to our presently observed cosmos. These considerations led several physicists, including the Russian Andrei D. Sakharov and Steven Weinberg, to propose in the 1960s the non-conservation of baryon number.

Despite all the difficulties, such experiments have been planned and performed in several countries since the early 1980s. One of the most elaborate systems consists of 8000 tonnes of water in a container which is encircled with 2000 photomultipliers sensitive to Cerenkov radiation and situated some 600 metres below the ground, in a salt mine near Cleveland, Ohio, in the USA. The system began operating in 1982.

Up to now no evidence has been recorded in the Cleveland experiment for the decay of the proton. This result is consistent with data from similar detectors the world over, and the experimental lower limit for the proton lifetime stands currently at about 10^{32} years. Despite these negative results, it is not yet established that the proton is stable. There are other theories which allot the proton a longer life span (though not much longer than 10^{33} years), and the sensitivity of the experiments will have to be improved before the question can be settled.

If the proton is found to be unstable, it means that the entire known universe has a finite life expectancy, because the protons in the nuclei of all the elements would at long last disintegrate. This gloomy forecast does not loom around the corner – 10^{32} years is an eternity, not only compared to human life and human history, but also compared to the age of the universe (10^{10} years according to our present knowledge). Yet it may cast some new light on our concepts of cosmic evolution.

Conservation of lepton numbers

A conservation law similar to the law of the conservation of baryon number exists for leptons as well. Electrons, muons and neutrinos always appear and vanish in particle–anti-particle pairs. For example, in the beta decay, $n \rightarrow p + e^- + \bar{\nu}_e$, a lepton ($e^-$) and an anti-lepton ($\bar{\nu}_e$) are created. (For this reason the neutrino in this process is defined as an anti-particle!) In pair production, the photon is converted into an electron–positron pair, and in the decay of a pion to a muon, an anti-neutrino is also created:

$$\pi^- \rightarrow \mu^- + \bar{\nu}_\mu$$

Until 1962, physicists believed that a law conserving 'lepton number' existed similar to the conservation law of baryon number, and that each lepton should be given a lepton number $L = 1$, and each anti-lepton a lepton number $L = -1$. It is easy to see that pair production and the decay of pions and muons are consistent with this rule, and actually no process was found to violate it. But the physicists were not yet satisfied, and that was because of the following phenomenon: the muon has never been observed to decay in the manner $\mu^- \not\rightarrow e^- + \gamma$, even though such an interaction does not seem to contradict a single conservation law (a lepton of electric charge -1 appears on each side of the equation). In 1962 it was proven that the neutrino of the electron differs from the neutrino of the muon. (The experiment is described in section 4.3.) It then became clear that leptons are governed by *two* separate conservation laws: a specific conservation law for the 'electron family', and a second conservation law for the 'muon family'. In the first family we find two particles (ν_e, e^-) possessing an 'electronic lepton number' $L_e = +1$, and two anti-particles ($\bar{\nu}_e$, e^+) for which $L_e = -1$. In the second family, μ^- and ν_μ have a 'muonic lepton number' $L_\mu = +1$, while for μ^+ and $\bar{\nu}_\mu$, $L_\mu = -1$. In each interaction, the electronic lepton number and the muonic lepton number must each be conserved. For this reason a muon cannot decay into an electron and a photon, but only as follows:

$$\mu^- \rightarrow e^- + \bar{\nu}_e + \nu_\mu$$

On each side the muonic lepton number is +1 and the total electronic lepton number is 0. The lepton numbers of the leptons we have encountered thus far are summarized in Table 7.1 (the lepton

Table 7.1. *Lepton numbers*

Particle	Electronic lepton number (L_e)	Muonic lepton number (L_μ)
e^-	+1	0
ν_e	+1	0
e^+	−1	0
$\bar{\nu}_e$	−1	0
μ^-	0	+1
ν_μ	0	+1
μ^+	0	−1
$\bar{\nu}_\mu$	0	−1

Table 7.1 depicts the situation in the 1960s, before the discovery of the heavy lepton described in section 10.7.

numbers of particles which are not leptons are zero of course). Later (Chapter 10) we shall meet a third family of leptons, with a similar 'tau lepton number' conservation law.

The symmetry principles responsible for the conservation of the lepton numbers and the baryon number are not yet clear. These 'charges' do not induce any fields, and the conservation laws they obey might be 'approximate' laws which are violated in some very rare cases.

The lepton and baryon number conservation laws govern the creation and destruction of all the known *fermions* (particles with half-integer spin). No similar law exists for bosons (particles possessing an integer spin). Mesons and photons, for example, are thus created and disappear without limitation (subject, of course, to the classical conservation laws of energy, charge, linear momentum and angular momentum).

7.7 The conservation of strangeness

Strangeness, like baryon number, is a sort of 'charge' which was defined, in an *ad hoc* manner, specifically for a conservation law based on experimental data. When physicists first began to study the kaons and hyperons (see Tables 5.1 and 5.2) they found that the probability of their creation in a collision between two protons of suitable energy is quite high. Calculations showed that such an interaction occurs whenever two protons are in contact for as little as 10^{-22} seconds, indicating that this process is a strong interaction. The life span of these particles, however, is of the order of 10^{-10} seconds, thus their decay is 10^{12} slower than a typical strong interaction. This fact, and additional evidence, clearly implied that the weak interaction is responsible for the decay of strange particles. This phenomenon presented a puzzle: until the discovery of strange particles, all experiments indicated that every particle decays in the quickest manner possible. And here we see particles which can be affected by the strong force (they were, indeed, *created* in a strong interaction) and decay to give other particles which are also influenced by the strong force and yet – their decay is by means of the slower, weak force. The lambda hyperon, for example, decays to give a nucleon and a pion, both of which participate in strong interactions. The charged kaon can decay in many ways, one of which is to three pions. The remainder of the strange particles decay into pions, nucleons and other strange particles. Why don't these decays occur within 10^{-22} seconds from the formation of the particle? What is the secret of the long life of these particles? Since no answer was at hand, these particles were awarded the nickname 'strange'.

This nickname had yet another justification. The American physicist A. Pais pointed out the fact that strange particles are generally created in pairs, and never as single particles. This fact was not evident when the study of strange particles was confined to cosmic rays, since only one member of the pair generally left its track in the cloud chamber. But in 1952, when the Cosmotron at Brookhaven was first employed, the formation of strange particles was researched in depth, proving this fact beyond all doubt. In characteristic processes of strange-particle formation, a lambda and a kaon are created together (see Fig. 7.15):

$$\pi^- + p \rightarrow \Lambda^0 + K^0$$

or a kaon and a sigma:

$$\pi^+ + p \rightarrow K^+ + \Sigma^+$$

or two kaons:

$$\bar{p} + p \rightarrow K^- + K^0 + \pi^+ + \pi^0$$

On the other hand, a reaction producing a single strange hadron such as:

$$p + p \not\rightarrow p + \Lambda^0 + \pi^+$$

was never observed, even if it does not contradict any of the conservation laws discussed so far.

Associated production is another 'strange' property of our strange particles; but it gives a hint which can help solve our puzzle. Let us recall the formation of the electron–positron pair from a γ photon ($\gamma \rightarrow e^+ + e^-$). The electromagnetic force is responsible for this process. Yet, after the two particles have been formed, neither

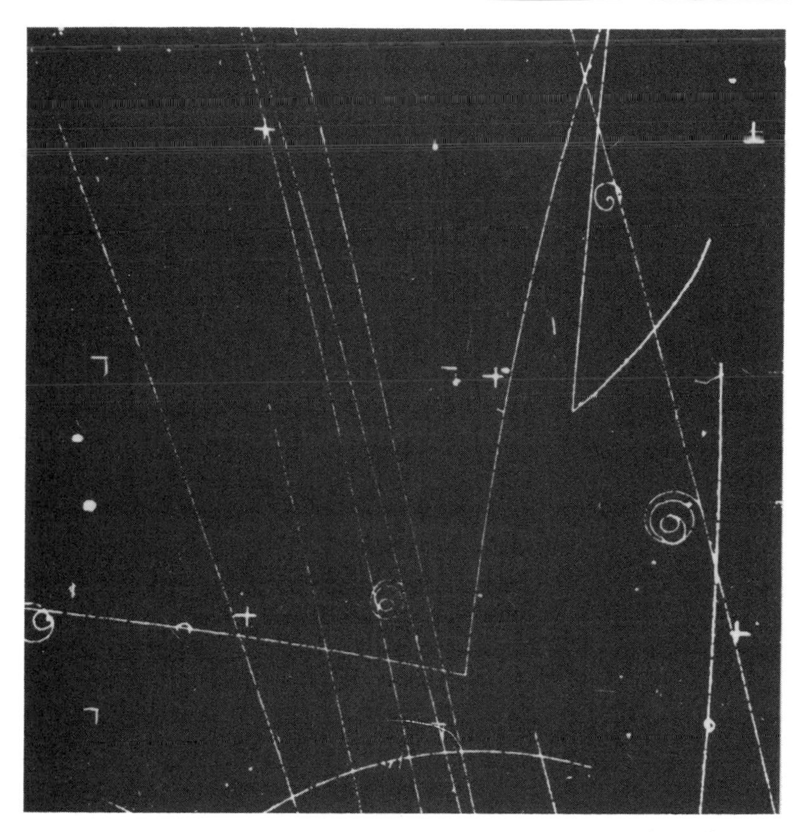

Figure 7.15. A bubble chamber photograph of the reaction $\pi^- + p \rightarrow \Lambda^0 + K^0$ which occurred when a π^- beam entered the hydrogen-filled chamber. The paths of the pions are the parallel straight lines drawn obliquely along the photograph. The paths of the Λ^0 and K^0 are not seen since these particles are electrically neutral, but they are identified through their decay products (see Fig. 7.16). (Courtesy Lawrence Berkeley Laboratory, The University of California, Berkeley, California.)

of them decays spontaneously, due to the law of charge conservation. The particles disappear only when they encounter their respective anti-particles.

Can a similar conservation law account for the decay of strange particles via the weak interactions? Indeed, as we described in Chapter 5, the very same idea occurred to M. Gell-Mann from the USA and K. Nishijima from Japan, who in 1953 proposed independently the following theory. A certain quantum number, which they termed 'strangeness' (S), exists which is conserved in strong and electromagnetic interactions but not in weak interactions. In other words, if a system of particles goes through a strong or electromagnetic interaction, the strangeness after the interaction must

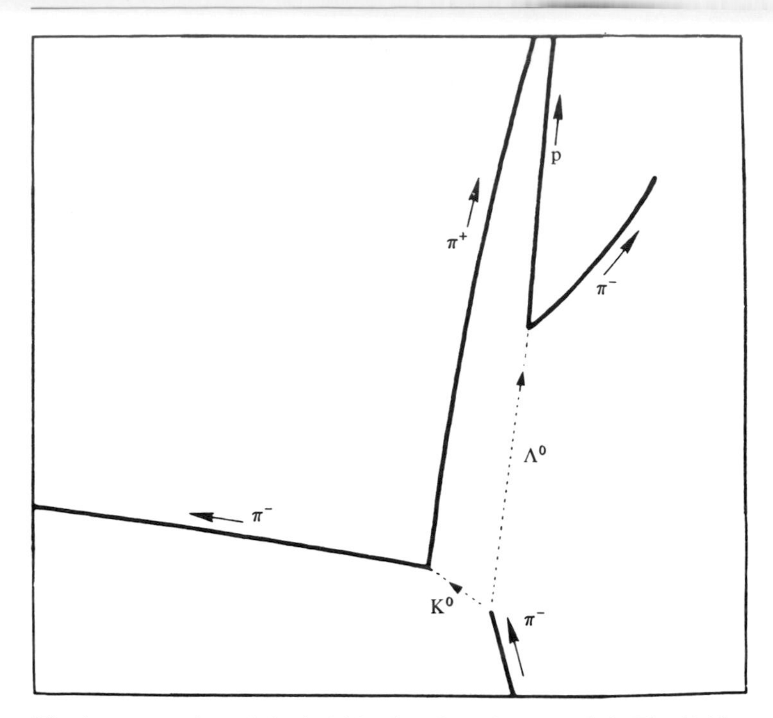

Figure 7.16. The interpretation of the bubble chamber photograph in Fig. 7.15.

equal the strangeness prior to the interaction. However, in a weak interaction the total strangeness of a system can change. By ascribing zero strangeness to the nucleons and pions, and non-zero strangeness to the kaon and hyperons, all the puzzles associated with the strange particles were immediately resolved.

When strange particles are formed through a strong interaction in a collision between two protons or between a proton and a meson, *the total strangeness of the products must equal zero*, since the strangeness of the original particles is zero. For this reason strange particles are created in groups of two or more. For example: a particle with strangeness of +1, along with a particle of strangeness −1:

$$\begin{array}{lcccc} & \pi^- + & p \rightarrow & \Lambda^0 + & K^0 \\ S = & 0 & 0 & -1 & 1 \end{array}$$

and the total strangeness is zero before the reaction as well as after it. Once a strange particle is created, it cannot decay in a strong or electromagnetic interaction to particles lacking strangeness or of lower strangeness than its own, because this will violate the strangeness conservation law. Nevertheless, it can decay to such particles via the weak interaction because the law of conservation of strangeness (as opposed to the other conservation laws we have met so far) does not apply to weak interactions! In weak processes strangeness can change. By analysing the reactions in which strange particles are produced Gell-Mann and Nishijima could deduce the amount of strangeness that should be ascribed to each particle. The Λ^0 was assigned – arbitrarily – a strangeness of -1, and by assuming that strangeness is conserved in the reaction $\pi^- + p \rightarrow \Lambda^0 + K^0$, that of K^0 was determined as $+1$. Similarly it was found that for K^+ too, $S = +1$, and that all the sigma hyperons (Σ^+, Σ^0, Σ^-) have a strangeness of -1. From the reactions in which the xi hyperons (Ξ^-, Ξ^0) are created their strangeness was found to be -2:

$$\begin{array}{rccccccc} & p & + p \rightarrow & \Xi^0 & + p & + K^0 & + K^+ \\ S = & 0 & 0 & -2 & 0 & +1 & +1 \end{array}$$

The omega minus, which is usually produced in accelerators in a collision between a kaon and a proton, should have a strangeness of $S = -3$, if strangeness is to be conserved in the process;

$$\begin{array}{rcccccc} & K^- & + p \rightarrow & K^0 & + K^+ & + \Omega^- \\ S = & -1 & 0 & +1 & +1 & -3 \end{array}$$

The strangeness quantum numbers of the kaons and hyperons are summarized in Table 7.2. Note the relation between the mass and strangeness of hyperons: the heavier the particle, the more negative its strangeness. The reason for this phenomenon was unexplained when strangeness was defined, but later it received a simple and surprising explanation, as we shall see.

The strangeness of the other particles appering in Tables 5.1 and 5.2 (the nucleons, pions and η^0) is zero. If the strangeness of a given particle is S, the strangeness of its anti-particle will be $-S$. Strangeness plays no role in interactions among leptons and photons, and is thus formally defined as zero for these particles.

The model of Gell-Mann and Nishijima explained both the creation of the strange particles in pairs, and their longer than expected lifetime. Investigation of the decay modes of the kaons and hyperons has shown that most of them can decay only by weak interactions, because the process cannot help but violate the conser-

Table 7.2. *The strangeness of the kaons and hyperons*

Particle	K^+	K^0	$\bar{K}^0$	K^-	Λ^0	Σ^+	Σ^0	Σ^-	Ξ^0	Ξ^-	Ω^-
Strangeness (S)	+1	+1	−1	−1	−1	−1	−1	−1	−2	−2	−3

vation of strangeness. The kaons are the lightest strange particles and therefore their decay products are all of zero strangeness. The decay products of the Λ^0, Σ^+ and Σ^- must include a nucleon because of the conservation of baryon number, and the remaining mass is too small for a kaon which could have taken the strangeness of the original particle, thus the decay of these particles, like the decay of the kaons, cannot conserve strangeness.

$$\begin{array}{lccccc} & \Lambda^0 & \xrightarrow{\hspace{3em}} & p & + & \pi^- \\ S = & -1 & \begin{pmatrix}\text{weak}\\ \text{interaction}\end{pmatrix} & 0 & & 0 \qquad (\Delta S = +1)\end{array}$$

The decay of the xi particles to particles of zero strangeness is a two-stage process, and in each stage the strangeness increases in one unit.

$$\begin{array}{lccc} & \Xi^0 \rightarrow & \Lambda^0 + & \pi^0 \\ S = & -2 & -1 & 0 \qquad (\Delta S = +1)\end{array}$$

$$\begin{array}{lccc} & \Lambda^0 \rightarrow & p + & \pi^- \\ S = & -1 & 0 & 0 \qquad (\Delta S = +1)\end{array}$$

Similarly, the omega minus transforms to non-strange particles in a three-stage process:

$$\begin{array}{l} \Omega^- \rightarrow \Xi^0 + \pi^- \\ \Xi^0 \rightarrow \Lambda^0 + \pi^0 \\ \Lambda^0 \rightarrow p + \pi^- \end{array}$$

where again the strangeness increases by one unit in each stage. It can be proved that in all these cases, the conservation laws of charge, energy–mass, and baryon number prevent the decay of the particles to particles of the same strangeness, so that the only possible route open to them is a decay which does not conserve strangeness, which necessarily is a weak interaction. The Σ^0 is the only hyperon which can decay without violating the conservation of strangeness:

$$\begin{array}{lccc} & \Sigma^0 \rightarrow & \Lambda^0 + & \gamma \\ S = & -1 & -1 & 0 \qquad (\Delta S = 0)\end{array}$$

and its lifetime is indeed shorter by several orders of magnitude than the lifetimes of the other hyperons. Yet, since the photon is not affected by the strong force, this decay is an electromagnetic rather than a strong interaction.

To sum up, each hadron can be assigned a type of 'charge' called strangeness, S, which is conserved in strong and electromagnetic interactions ($\Delta S = 0$) but not in weak interactions (where $\Delta S = \pm 1$ or 0). The law of the conservation of strangeness explains why the strange particles are produced in pairs, and why their decay is so slow.

We have seen that every hadron is characterized by three types of 'charge': an electric charge Q, a baryon number (or charge) A, and a strange charge or strangeness S. Each of these charges is a whole number – positive, negative or zero – and the charges of the respective anti-particle are opposite in sign to those of the particle. If a hadron has the 'charges' Q, A and S, the charges on its anti-particle will be $-Q$, $-A$ and $-S$. The conservation laws which apply to these charges are simple counting rules, based on experimental facts, such as the finding that certain processes have not been observed although they do not violate the classical conservation laws.

Strangeness and the neutral kaons

Another mystery which was solved thanks to the discovery of strangeness was connected to the neutral kaons. It was found that a beam of neutral kaons produced in an accelerator usually consisted of two types of particles. Both types possess almost the same mass (497.7 MeV) but they differ in their life span and mode of decay. The particle with the shorter life, which will be denoted by K^0_S (S for short), decays after 8.9×10^{-11} seconds, on average, to two pions. The average lifetime of the other neutral kaon, which will be denoted by K^0_L (L for long), is longer – 5.2×10^{-8} seconds – and its usual mode of decay is to three particles (see Table 5.1). If the beam is allowed to travel a certain distance, almost all the short-lived K^0_S will die out and the beam will turn into an almost pure K^0_L beam. Therefore at the beginning the decay to two particles is dominant, while after travelling some distance most of the decays are into three particles.

It was found, in a brilliant experiment, that there is a tiny difference between the masses of the two particles: the mass of the K^0_L is greater by 3.6×10^{-12} MeV than that of the K^0_S. Because of the differences in lifetimes, decay modes and masses, it is clear that K^0_L and K^0_S are not a particle–anti-particle pair. What then is the relationship between them?

In 1955 Gell-Mann and Pais provided an amazing answer to this question. They suggested a theory which shed a new light on the neutral kaons, making them different from any other neutral particle we have met so far. (Actually most of the experimental research into the properties of the neutral kaons was carried out after this theory had been published and was initiated by its predictions. One might say that the theory provided the answers even before the questions were posed.) According to this theory there is a neutral kaon, K^0, which has an anti-particle $\bar{K}^0$. Both have exactly the same mass, spin (0) and electrical charge (0). They differ, however, in their strangeness. K^0 possesses positive strangeness ($S = +1$) while its anti-particle, $\bar{K}^0$, possesses, of course, negative strangeness ($S = -1$). The most rapid decay mode of each is to two pions ($\pi^+\pi^-$ or $\pi^0\pi^0$) by a weak interaction in which the strangeness is changed by one unit.

Here the wavelike nature of material particles, introduced by de Broglie, comes into action. In a beam which contains K^0 and $\bar{K}^0$, interference occurs between the wave functions of the particle and its anti-particle, in the same way as two beams of light interfere with each other. In quantum mechanics such an interference can be displayed as a sum or difference of two wave functions (similar in a way to the constructive and destructive interference of light waves). At points where the wave functions are added together, the decay is to two pions, which is a relatively fast process. However, at points where the wavelengths are subtracted, the decay into two pions is forbidden because of a conservation law of a quantity called *CP* parity (see section 7.9). In this case the only decay mode available is into three particles, and such a process is always more complicated and slower than the two-particle decay. The observed behaviour of the kaon beam is as if it contains two types of particles: fast-decaying particles which are the K^0_S and slow-decaying particles – the K^0_L.

It can be shown that this theory explains the observed difference between the masses of K^0_L and K^0_S. Another remarkable consequence of the theory which has been confirmed by experiments is: both K^0_L and K^0_S possess 'mixed' strangeness (+1 and −1) since both are combinations of K^0 and $\bar{K}^0$. Therefore, each of these 'particles' can participate in processes which require either positive or negative strangeness. As a matter of fact the combination states K^0_L and K^0_S can be treated as real particles in every respect. The description of the kaon beam as composed of K^0_L and K^0_S is as legitimate as its description in terms of K^0 and $\bar{K}^0$.

Strangeness and multiplets of particles

Because of the small mass difference between the proton and the neutron, which are quite similar in several other aspects (they possess the same spin, strangeness and baryon number), they may be considered as two states of one particle (nucleon), which can appear with or without electric charge. Looking at Tables 5.1 and

5.2 you will find that other hadrons too can be arranged in groups, so that the particles in each group have the same strangeness, baryon number and spin, and only slightly different masses. These groups are called multiplets. Among the hadrons we have met there are doublets (K^+, K^0), ($\bar{K}^0$, K^-), (p, n), (Ξ^0, Ξ^-), triplets (π^+, π^0, π^-), (Σ^+, Σ^0, Σ^-) and singlets: η^0, Λ^0, Ω^-. For every multiplet of baryons there is a similar multiplet of anti-baryons. (Note that the three pions are arranged in one multiplet while the four kaons are divided into two separate multiplets. This is because the three pions have the same strangeness, while in the kaon family the strangeness of the K^+K^0 differs from that of $\bar{K}^0K^-$. Only particles of the same strangeness are included in the same multiplet.)

The various multiplets differ from each other not only in number of particles but in average charge as well. In the nucleon doublet, for example, the average charge is $+\frac{1}{2}$ (1 for the proton and 0 for the neutron), while that of the $\Xi^0\Xi^-$ doublet is $-\frac{1}{2}$. In the next section we shall see that by arranging the hadrons in multiplets, two new conservation laws can be formulated. But first we shall point out an interesting relationship between strangeness and the average charge of the multiplet, which was noticed by Gell-Mann and Nishijima as soon as they defined strangeness, and even before the whole gang of strange particles was discovered.

The average charge of the nucleon doublet (p, n) is $\frac{1}{2}$ and their strangeness is zero. The average charge of the Σ triplet as well as the Λ singlet is 0, and the strangeness in both cases is -1. The xi particles (Ξ^-, Ξ^0) have an average charge of $-\frac{1}{2}$ and strangeness of -2, while the charge of the omega minus is -1 and its strangeness -3. One can immediately see an obvious relation between average electric charge and strangeness: the more negative the average charge, the more negative the strangeness. A similar relation holds for mesons as well. Actually by defining a new quantum number (hypercharge) as the average charge of a multiplet, multiplied by two, this relation can be summarized in one simple equation:

$$Y = S + A$$

where Y is the hypercharge, S the strangeness and A the baryon number. This equation is valid for all hadrons: baryons, anti-baryons and mesons.

This unison between strangeness and the average charge of the multiplet is quite surprising, when we recall that the original definition of strangeness had nothing to do with either electric charge or multiplets. This implies that the quantum number which was artificially 'invented' in order to explain the 'strangeness' of strange particles has, in fact, more significance than first meets the eye.

When Murray Gell-Mann and Kazuhiko Nishijima defined strangeness and discovered the relationship between it and the average charge of the multiplets, not all the hyperons had yet been found. The Ξ^0 was not yet known, although Ξ^- had already been discovered, and its strangeness found to be -2. The two physicists concluded that Ξ^0 must also exist, since by substituting $S = -2$ and $A = 1$ in the equation above, they got $Y = -1$, or $-\frac{1}{2}$ for the average charge of the multiplet! The Ξ^0 was finally discovered, by detecting its decay products (since it is a neutral particle, it does not leave tracks in the bubble chamber).

Among the sigma particles, Σ^- and Σ^+ were then known, but not the short-lived Σ^0. Moreover, its existence could not have been predicted by the -1 strangeness of the sigma particles (leading to $Y = 0$) since even without Σ^0 the average charge of the multiplet is zero! In spite of this, Gell-Mann and Nishijima predicted that a neutral sigma particle does exist, after analysing the properties of the quantum number known as 'isospin', which you will read about in the next section. Another impressive victory was the prediction of the existence of Ω^-, which we will read about later on.

7.8 The conservation of isospin and I_3

By arranging the hadrons in multiplets, two additional quantum numbers could be defined, the first of which is conserved only in strong interactions, while the second is conserved both in strong and electromagnetic interactions.

The isospin is also called I-spin.

The first of these quantum numbers is called *isotopic spin* or *isospin* and denoted by I. Despite its name it has nothing to do with ordinary spin or with any other sort of angular momentum (although its definition bears some mathematical likeness to spin). As opposed to strangeness, which is related to the *average charge* of the multiplet, the isospin is related to the *number of particles* in the multiplet. If there are n particles in a multiplet, the isospin (I) of each of the particles in this multiplet is:

$$I = \frac{n-1}{2}$$

The isospin of the solitary particles, η^0, Λ^0, Ω^-, is therefore zero. Each of the doublets (p, n; K^+K^0; $\Xi^0\Xi^-$) possesses isospin of $\frac{1}{2}$, and the triplets (pions and sigmas) isospin of 1. The isospin of each particle equals that of its anti-particle.

The isospin was defined by Heisenberg in the 1930s in order to

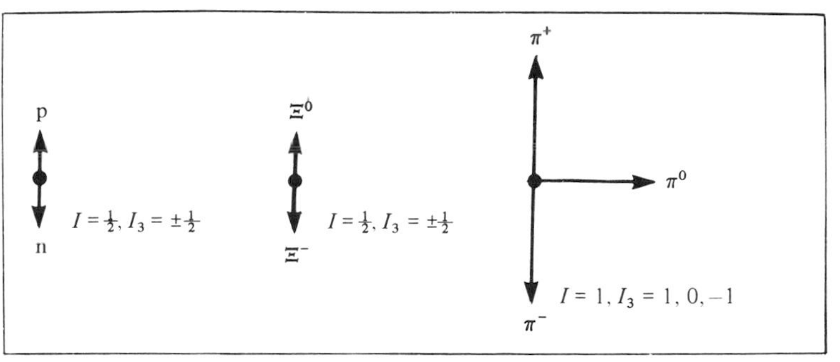

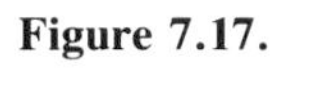
Figure 7.17. The isospin, I, and its abstract projection, I_3, were defined in analogy to the spin and its projection along a direction in real space. I_3 can take $2I + 1$ values, between I and $-I$.

characterize the nuclides of various isotopes (therefore its name). It may seem easier to define it simply as the number of particles in the multiplet, but physicists chose a slightly more complicated definition to make it easier to formulate the relevant conservation law. The definition, indeed the very name 'isotopic spin', stems from an analogy, slightly artificial albeit, between the charge and spin of particles. Recall that a spin of magnitude J (in units of $h/2\pi$) can orient itself in $2J + 1$ different states with respect to a defined direction in space. Heisenberg assumed that similar particles such as the neutron and proton are two different states of one basic particle, like the two spin states of the electron. He therefore denoted these states by the direction of an imaginary vector with properties similar to the properties of the spin. To have two states, that vector must have a length of $\frac{1}{2}$; to obtain three states, a length of 1 and so on. This vector is the isospin. We must remember, however, that the isospin vector is just a convenient abstract notation, and has nothing to do with a real spin.

To distinguish between different particles in the same multiplet, each is assigned a quantum number called I_z or I_3 which is the projection of the isospin vector along the z-axis in the imaginary space in which it is defined. The proton, for instance, is assigned $I_3 = \frac{1}{2}$ (or 'upwards' isospin) and the neutron $I_3 = -\frac{1}{2}$ ('downwards' isospin). I_3 of each pion is equal to the charge (see Fig. 7.17). For all the singlets $I = I_3 = 0$.

The isospin is conserved in strong interactions, i.e. the total isospin of the particles participating in a strong interaction equals the total isospin of the products. I_3, on the other hand, is conserved in both strong and electromagnetic interactions. The conservation law of I_3 is a simple counting rule while that of the isospin I is more complicated, since it is a vector quantity, and it is usually possible to add the isospins of several particles in more than one way. It should be mentioned that both I and I_3 are meaningful for hadrons only.

I-spin is like angular momentum, albeit in an abstract space. I-spin conservation corresponds to invariance under rotations in that imaginary space (I-space), just as the conservation of angular momentum reflects rotational invariance in the real space. This means that the strong interactions do not have a preferred direction in I-space, they do not distinguish between 'up' and 'down', between a proton and a neutron. I-spin invariance thus explains the fact that it takes about the same energy to extract a proton or a neutron from a nucleus. The forces between two protons, two neutrons or a neutron and a proton are the same (except for the electromagnetic contribution which is very small compared to the strong interaction). This is another way of saying that the strong force is not affected by charge, and acts identically on all the particles in a given multiplet. Actually, this is the experimentally identified principle of symmetry from which the conservation law of isospin is derived, and this also explains why the isospin is not conserved in electromagnetic and weak interactions which do depend on charge.

There is an interesting relationship between I_3 and other quantum numbers. It can be shown that the charge Q of every hadron (in units of the proton charge) is given by the equation:

$$Q = I_3 + \tfrac{1}{2}(A + S)$$

This equation helped physicists conclude that Σ^0 existed before it was discovered experimentally (when the strangeness of its charged brothers was determined, it turned out that their I_3 values are ± 1, and it became clear that the Σs are a trio rather than a couple). From this equation it is also evident that I_3 is conserved in every interaction in which S is conserved (since A and Q are conserved in every interaction).

7.9 The TCP theorem

In the previous sections, the kinship of the conservation laws to symmetry principles or to the invariance of the basic interactions under certain operations was repeatedly mentioned. In this section three more conservation laws will be described, but this time we shall do it in the reverse order, and present the symmetry principles first. Let us assume that we can observe a certain reaction between particles and can even film the event. Let us take, for example, the reaction

$$\mathrm{A} + \mathrm{B} \rightarrow \mathrm{C} + \mathrm{D}$$

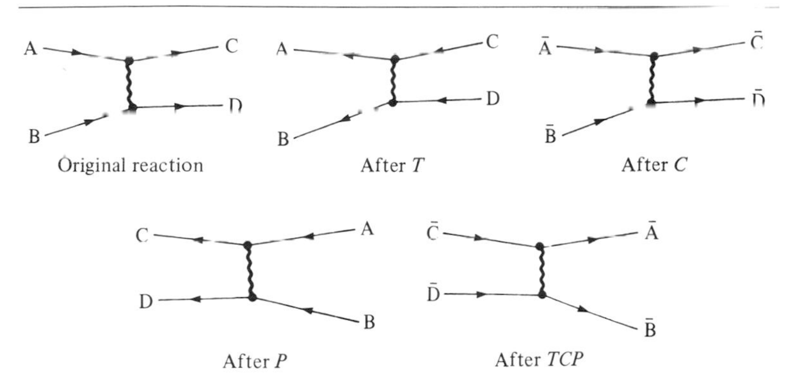

Figure 7.18. A schematic description of the effect of the operators T, C and P on the reaction A + B → C + D.

which is schematically depicted in Fig. 7.18. The particles A and B collide and in the course of this collision the particles C and D are produced in their place. We have seen that various conservation laws stem from the fact that this reaction could take place in exactly the same manner at any point in space, at another time, or with the original particle moving in other directions. In other words: transforming the reaction in time or in space will not lead to some impossible event which would cause us to raise our eyebrows in wonder.

We can ask if there are any other operations which we could carry out, in our imagination, on this reaction and then check to see if we obtain possible results. Physicists concluded that three basic operations of this kind exist, and named them T, C and P.

T indicates the time-reversal operation. The intention here is a sheer mental exercise (no time machine is implied). Let us assume that we have filmed the reaction and we are now viewing the film in reverse. We will see the particles C and D colliding and disappearing and the particles A and B suddenly appearing (see Fig. 7.18). Is such a reverse reaction possible? Might it possibly occur in nature or in an accelerator? If the answer is affirmative, we can state that the original reaction (and its reverse, of course) is *invariant* – or non-changing – with respect to the operation T. However, if the reverse of the film leads to a reaction which is impossible for some reason, the original reaction is not invariant with respect to T.

P is a simple operation known as reflection or inversion. (P stands for *parity*.) All you have to assume here is that you are watching the interaction in a mirror (more precisely, P is a reflection in the mirror and an upside-down turning over. This operation transforms a point whose coordinates are (x, y, z) to the point $(-x, -y, -z)$). Here

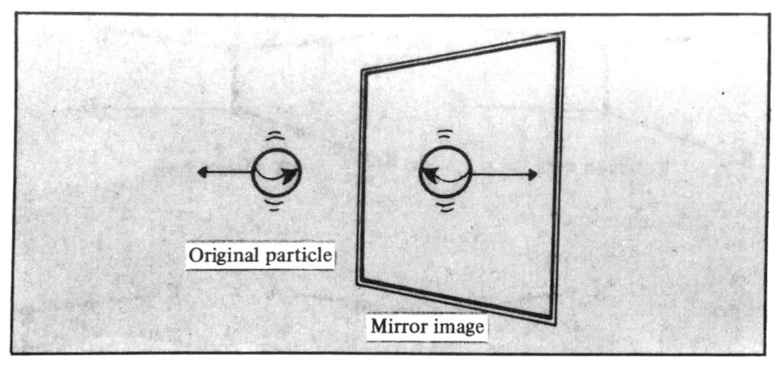

Figure 7.19. The reflection in a mirror reverses both the direction of motion and the direction of the spin. To see this you have to imagine that the particle actually spins on its axis. In the figure the original particle travels to the left, and spins counter-clockwise ('spin up', according to the rule of the right-handed screw). In the mirror the particle travels to the right, and spins clockwise ('spin down').

again we ask the question – is the reaction in the mirror possible or not? Note that this operation affects not only the direction of motion of a particle, but also its spin (in this case the particle may be considered actually to spin on its axis, see Fig. 7.19).

C is an operation called 'charge conjugation': all you have to do here is to turn each particle participating in the reaction into its own anti-particle, again the purpose being to see whether or not the reaction we obtain is possible. By operating *C* on the reaction $\mathrm{A} + \mathrm{B} \rightarrow \mathrm{C} + \mathrm{D}$ the reaction $\bar{\mathrm{A}} + \bar{\mathrm{B}} \rightarrow \bar{\mathrm{C}} + \bar{\mathrm{D}}$ is obtained.

During the 1950s, a number of physicists proved that if we apply all three operations *T, C* and *P* (in any order) on an equation depicting any reaction between particles, we obtain a reaction which is also possible. In other words, all interactions – strong, electromagnetic and weak – are invariant with respect to the combined operation *TCP*. This theory, known as the '*TCP* theorem' has successfully stood the test of numerous experiments.

And what about each of the three operations exerted separately? Here we have one of the most thrilling stories of particle physics. Up until the 1950s physicists thought that the invariance principle held for each of the three operations separately. There was no theoretical proof for this, but examination of various typical reactions gave the impression that application of *T, C* or *P* on any reaction between particles would provide a possible process. Let us have a look, for example, at beta decay: $\mathrm{n} \rightarrow \mathrm{p} + \mathrm{e}^- + \bar{\nu}_\mathrm{e}$. Application of *C* to this weak interaction gives rise to the process $\bar{\mathrm{n}} \rightarrow \bar{\mathrm{p}} + \mathrm{e}^+ + \nu_\mathrm{e}$ which is a process that has been observed experimentally. Application of T to the original reaction results in a more problematic process: $p + e^- + \bar{\nu}_e \rightarrow n$. The probability of observing such a process experimentally is negligible since it necessitates a collision between three particles (the application of T on macroscopic phenomena results in processes whose probability is just as low. Think, for example, of a parachutist taking off from the ground towards the plane, or of a hammer blow which converts shattered fragments of a vase into an unbroken one). At any rate, different pieces of evidence indicate that were we able to bring a proton, electron and anti-neutrino together in the desired manner, the interaction would indeed be possible. (Indeed, a similar process, $p + \bar{\nu}_e \rightarrow e^+ + n$, has been observed in the laboratory, as we saw in Chapter 3.)

And what about the operation P, or the reflection in a mirror? This seemed to be the most harmless of all three operations since the original reaction is hardly changed at all, and the assumption was that all interactions are invariant with respect to P, as well.

In 1956, physicists the world over were astounded when it was proven that the 'innocent' operation P does not yield an invariance principle. It was shown that if we look closely at the 'mirror image' of a beta decay, we will observe an impossible detail. In other words, this process, and other weak interactions, are not invariant under P. Later, it was shown that weak interactions are not invariant under C, either. The process $\bar{n} \rightarrow \bar{p} + e^+ + \nu_e$ does exist but, as we shall see below, the natural process differs somewhat from that which we would obtain by applying C on the process $n \rightarrow p + e^- + \bar{\nu}_e$.

We have mentioned the relationship between the conservation laws and symmetries. Up until now we have been dealing with only the symmetry (or invariance) principles of the operations C, P and T, and not with the conservation laws which result from them. In the next section, the conservation of parity is discussed more thoroughly.

7.10 The conservation of parity

This conservation law results from the invariance of certain interactions with respect to the inversion or reflection operation (P). The physical quantity of *parity* is not even defined in classical physics since it is related to the properties of wave functions. In order to determine what is the parity of a particle or of a system of particles, we must ask ourselves how the wave function will appear if reflected in the mirror. There are wave functions whose form is not at all affected by reflection. Those are wave functions which fulfil the equation

$$f(-x, -y, -z) = f(x, y, z)$$

where x, y and z are the coordinates in space and f is the value of the wave function at the given point. Such functions are called *symmetric* (or 'even') and are assigned a positive parity of +1. The wave function of the electron in the hydrogen atom in its ground state has spherical symmetry and is an example of a symmetric wave function. Some functions change sign under reflection, i.e.

$$f(-x, -y, -z) = -f(x, y, z)$$

and are called *antisymmetric* (or odd). Such functions are assigned a negative parity of −1. In both cases we say that the function has a well-defined parity. For some functions, $f(-x, -y, -z)$ equals neither $f(x, y, z)$ nor $-f(x, y, z)$. Such functions are said not to possess definite parity. We should point out here that every simple system – such as a solitary particle – may be described by a wave function of definite parity. Examples of one-dimensional functions possessing parities of +1 and −1 as well as functions without definite parity are illustrated in Fig. 7.20.

Note that in contrast to other quantum numbers, parity has only two values: +1 and −1. It was found that the parity of a system of two particles is the product of the parities of the two wave functions, provided that the angular momentum of the system is zero (otherwise the calculation is slightly more complicated). In fact, the terms positive and negative parity and the values of +1 and −1 were specifically chosen because of the following multiplication law:

$$(+1) \times (+1) = (-1) \times (-1) = +1; \ (+1) \times (-1) = -1$$

During the years 1954–56 physicists were preoccupied with a problem which was called the 'theta–tau riddle' (θ–τ). This problem appeared after it was proven beyond doubt that the θ meson, which decays into two pions, and the τ meson, which decays to give three pions, are one and the same particle (later named K meson or kaon). What perplexed physicists was not the fact that the particle decays in two alternative ways, but that this phenomenon violates the law of conservation of parity. And why so? In choosing wave functions to describe a particle, there is a certain amount of leeway, but since physicists had decided to characterize the proton by a wave function of +1 parity, an analysis of certain reactions teaches us that we must assign a parity value of −1 to all the pions. A careful analysis of the decay processes of the kaon into pions showed that when the decay produces two pions, the overall parity of the products is *positive*, while a decay into three pions results in overall

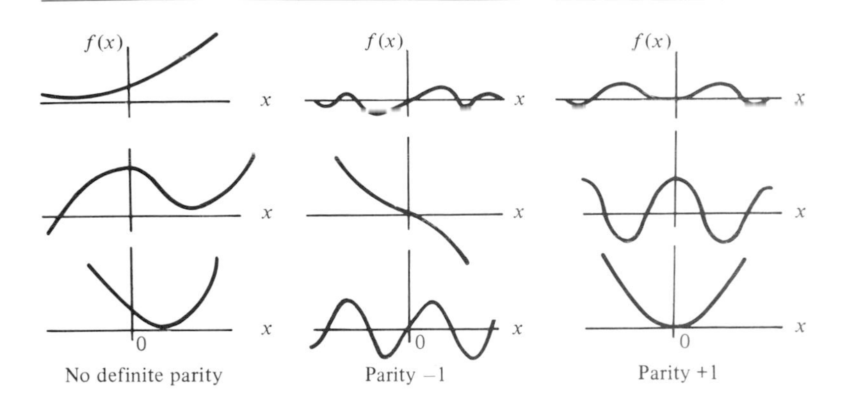

Figure 7.20. Examples of one-dimensional functions possessing parities of +1 and −1, and functions with no definite parity.

negative parity of the products. This was quite surprising. If the conservation of parity holds, then the parity of the kaon must be identical in each case to the overall parity of its decay products, yet how can its parity be positive and negative at one and the same time? You might suggest that we define the parity of the pion as +1, but this would not be of any help to us – in this case too the parity of the three pions produced by the kaon decay would be −1 (because of the angular momentum of the system).

The solution to this puzzle turned out to be a simple matter: parity is not conserved in every interaction. A lot of courage was needed, however, to stand up and say such a thing, since the law of conservation of parity was considered almost as unassailable as the law of energy conservation. Because of the relationship between conservation laws and symmetry principles, lack of parity conservation in certain interactions means that the mirror images of those interactions are not possible processes, which seemed preposterous. Moreover, since the mirror image interchanges right and left, the conservation of parity, or the invariance of the basic interactions with respect to P, means that *nature does not differentiate between right and left*, and this seemed, up to the 1950s, a self-understood truth.

Indeed, during the 1950s, physicists were certain that we could never explain to the inhabitants of a distant galaxy by radio transmission which side we define as left and which as right (or which is a left-handed screw and which a right-handed screw, or, for that matter, which direction of revolution is 'clockwise'. All these definitions are related to one another and one of them can help define the others.) The reason for this was that no physical process

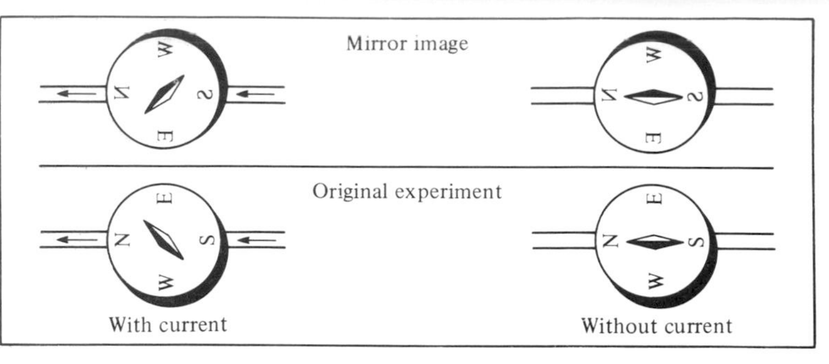

Figure 7.21. A compass on a conducting wire, as viewed from above. When the wire carries current in the northern direction, the north pole of the needle is deflected to the right, while in the mirror image of the experiment it is deflected to the left.

was known to prefer right to left! It was thought that every known process which takes place 'to the right' has exactly the same probability of occurring 'to the left', and that all physical processes look absolutely natural in the mirror.

Right and left, north and south

Here the astute reader who knows some physics may stop and ask: just one moment, I recall that when dealing with magnets, there is a difference between left and right. For example: when we place a piece of metallic wire under a compass, parallel to the magnetic needle, and run a current through the wire in a northern direction, we find that the magnetic field created by the current deflects the needle *to the right*. If we examine the same experiment in the mirror, the current will be in the same direction but the needle will be deflected *to the left*! (see Fig. 7.21). Have we not found here a process whose mirror image is not possible? However, if we study this example closely, we will find that full symmetry is conserved in this case as well. The magnetism in the magnet is the result of tiny microscopic electrical currents within matter which stem from the electrons orbiting the atom and the spins of electrons and nuclei. The reflection in a mirror reverses the direction of these currents as well (see Fig. 7.22). Therefore the application of the operation P on the magnet interchanges the north and south poles. If we consider this phenomenon, we will conclude that the actual mirror image of this experiment describes a possible situation.

We must further point out that we can tell which is the north pole and which is the south pole of a magnet with the help of the magnetic field of the earth, yet it is just by coincidence that the earth has

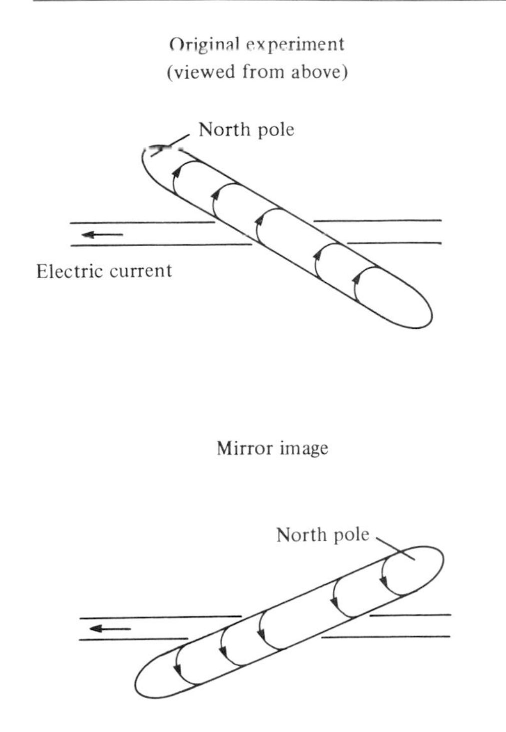

Figure 7.22. If we take into account the fact that the reflection in a mirror reverses the direction of the atomic electric currents which give rise to the magnetism of the compass needle, we find that the reflection also interchanges the north and south poles. (The presentation of the microscopic currents is very schematic.)

magnetic poles and that those poles are in the proximity of the geographical poles (it is believed that during various periods in the past their position was completely different from their present location). If we want to explain by radio broadcast to our friends in a distant galaxy which end of the magnet we call north, we will run into the same difficulties which we encountered when we wanted to explain to them which side is left and which is right, and what is clockwise. As Richard Feynman once joked, there are no little hairs on the north pole of a magnet.

A precise analysis of other physical processes always led to the same conclusion – nature does not distinguish between right and left and the 'mirror image' of any physical process is always possible, if we only know how to interpret it correctly. Under such circumstances, it would really require considerable courage to get up and announce that all the examinations and analyses were not precise enough and that in spite of everything, there are processes whose mirror images are not possible, and in which parity is thus not conserved. In fact, Richard Feynman and Martin Block raised the possibility at a conference in New York in 1956. Soon afterwards two young Chinese-born American physicists, 29-year-old Tsung-Dao Lee from Columbia University, and 33-year-old Chen Ning Yang from Princeton, made up a thorough analysis of all known facts, and concluded that indeed parity might not be conserved in weak interactions. Since they prepared their 'heresy' with clear logic, published it in the form of a convincing scientific paper, and even proposed some experimental methods of testing their hypothesis, and because an experiment which took place a few months later dramatically confirmed their argument, reward was not long in coming. The two scientists shared the 1957 Nobel prize for physics.

Let's examine, said Lee and Yang, the basis for the belief that all interactions conserve parity or are invariant with respect to reflection in a mirror. Many physical processes were carefully enough examined to ensure that their 'mirror images' do not contradict any natural law (sometimes a mental examination, similar to the one which we performed on the compass problem, is sufficient), but if we check these processes one by one we will find that they all result from either strong or electromagnetic forces. Since most of the physical phenomena that we encounter in everyday life and in the laboratory stem from these two forces, we are used to the notion that *every* mirror image of the world is natural and possible. But we must be very careful here! Weak interactions were studied much less intensively, and not with sufficient care. There is no proof – said Lee and Yang – that they also conserve parity. And since the decay of kaons is a weak interaction, perhaps this provides the answer to the θ–τ riddle.

The mirror distorts weak interactions

Lee and Yang even proposed a way of checking their hypothesis. They suggested choosing a weak process, beta decay from a radioactive atom, for example, which is a process which requires neither complicated equipment nor a costly accelerator, and examining the process carefully. Since they themselves were theoretical, not experimental physicists, they did not carry out the actual experiment. It was performed by a group of physicists from Columbia University and from the National Bureau of Standards in Washington DC, headed by another American-Chinese physicist – a woman this time – Chien Shung Wu.

In this experiment a piece of cobalt 60 (a beta emitter) was cooled

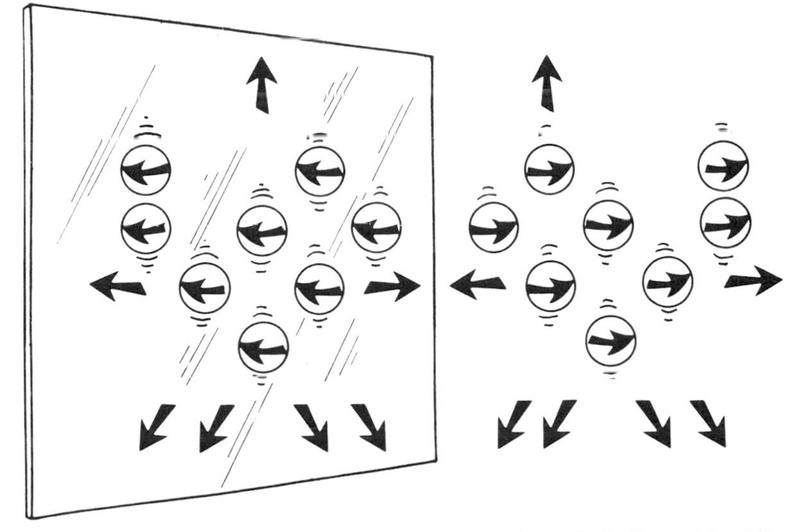

Figure 7.23. In Wu's experiment it was found that beta-emitting cobalt 60 nuclei, with directed magnetic moments, eject the electrons mostly in the opposite direction to that of the magnetic moments. (The spheres in the figures represent cobalt 60 nuclei, the magnetic moments of which point upwards. The arrows on the spheres indicate the direction of electric current which would produce such a magnetic moment. The other arrows mark the directions of the emitted electrons.) In the mirror image, on the other hand, most of the electrons are emitted in the direction of the magnetic moments, which now point downwards.

to a temperature very close to absolute zero and placed in a strong magnetic field. As a result a large portion of the cobalt nuclei became oriented so that their magnetic moments were in the direction of the field (at high temperatures such a feat would have been impossible, since the atoms possess thermal energy and oscillate in various directions, even when under the influence of a magnetic field). Now the researchers could check the directions in which the beta particles (the electrons) were emitted. And here awaited the great surprise: it was found that most of the beta particles were emitted in the opposite direction to that of the oriented magnetic moments! The mirror image of this process is not compatible with what happens in nature, since in the 'mirror' most of the particles are emitted in the direction of the magnetic moments (see Fig. 7.23). Just to give you an idea of the atmosphere of doubt which Lee and Yang's idea encountered before it was confirmed by experiment, W. Pauli was so skeptical that he wrote a letter prior to the experiment to V. Weisskopf stating that he was convinced the experiment would fail and that Lee and Yang were wrong. The

Pauli wrote in his letter: 'I do not believe that the Lord is a weak left-hander, and I am ready to bet a very high sum that the experiment will give a symmetric angular distribution of the electrons.'

experiment, however, was fully successful and proved that weak interactions are not invariant with respect to P; thus, parity is not necessarily conserved in these interactions. Nature does differentiate between right and left, and if we think hard, we can, with the help of the cobalt emission experiment, radio a distant galaxy the definitions of right and left (or alternatively, what direction is clockwise, or which is the north pole of a magnet). The conceptual revolution caused by this experiment is summarized in the following scheme.

The common belief prior to Wu's experiment was that:

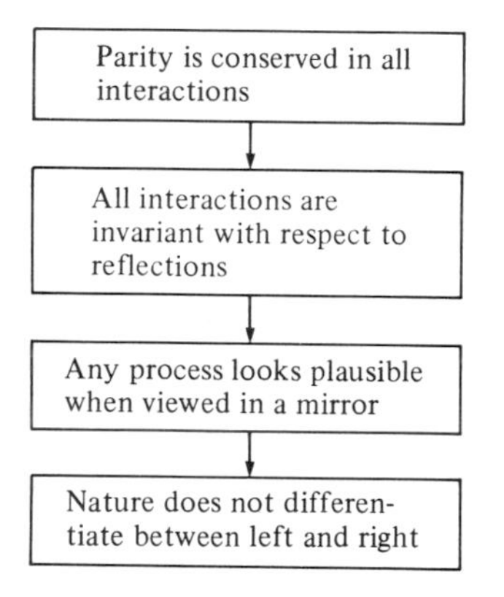

Wu's experiment proved that:

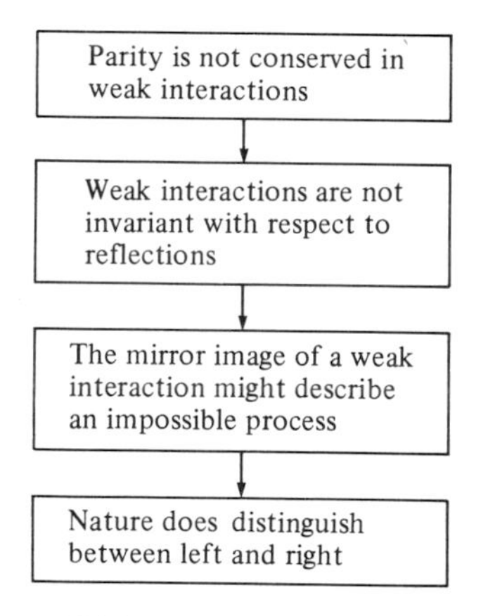

Wu's experiment solved the θ–τ puzzle. The parity of the kaon is today known to be -1, and its decay into two pions does violate the conservation of parity.

In 1978 Wu received the Wolf prize in physics in the Knesset (the Israeli Parliament) in Jerusalem. It was the first year of that international prize, and she was found worthy of being among the first winners.

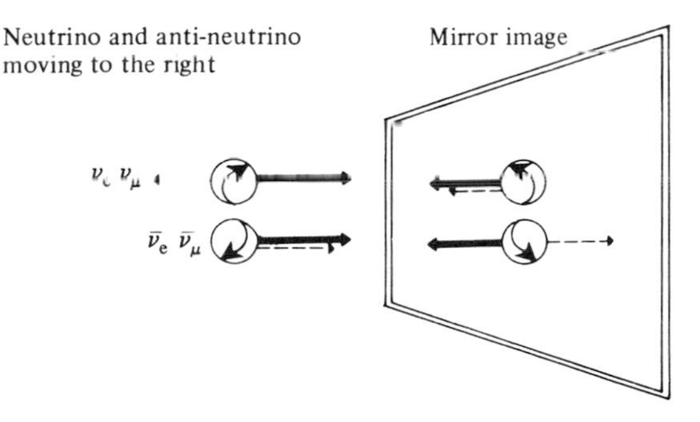

Figure 7.24. The spin of the neutrino is in the opposite direction to its motion, as in the case of a left-handed screw. The anti-neutrino, on the other hand, moves with its spin in the direction of its motion, as a right-handed screw. The mirror image, in both cases, is unrealistic. (The thick arrows indicate the directions of motion and the thin dashed arrows the spins.)

Over the course of time the non-conservation of parity was demonstrated in other weak interactions as well. For example, it was found that when the lambda decays into a pion and a proton ($\Lambda^0 \rightarrow \pi^- + p$), the emitted pion 'prefers' to move in the direction indicated by the spin of the Λ^0. (Here again, the mirror image will show the opposite preference, and therefore will be unrealistic.) The clearest proof, however, to the non-conservation of parity in weak interactions was provided upon the discovery that the neutrino, every neutrino, always moves with its spin in an opposite direction to that of its motion, while the anti-neutrino moves with its spin in the direction of its motion. Every process in which the neutrino participates, whether it is a beta decay or a decay of the pion into a muon, etc., will thus appear 'unnatural' in the mirror, since the mirror reverses the direction of spin (see the Fig. 7.24). As opposed to our previous conjecture, the application of the 'harmless' operation P on the process $n \rightarrow p + e + \bar{\nu}_e$ will result in an impossible process, since it changes an anti-neutrino that moves in a right-handed screw motion (which we denote $\bar{\nu}_{er}$) to one which moves in a left-handed screw motion ($\bar{\nu}_{el}$), and such a thing does not exist in nature.

$$P(n \rightarrow p + e + \bar{\nu}_{er}) = n \rightarrow p + e + \bar{\nu}_{el}$$

Moreover, the application of the operator C (charge conjugation) on a weak interaction will result in an impossible process as well, since C transforms a particle into its anti-particle *without changing the directions of its motion and spin*:

$$C(n \rightarrow p + e + \bar{\nu}_{er}) = \bar{n} \rightarrow \bar{p} + e^+ + \nu_{er}$$

And a ν_{er} (a neutrino whose motion resembles a right-handed screw) does not exist in nature! Without straining ourselves, and without even performing a single experiment we have proven that the weak interaction violates the invariance of the operation C, as well!

And what about the combined operation CP? If we apply it to the above process we will obtain a possible reaction:

$$CP(\mathrm{n} \rightarrow \mathrm{p} + \mathrm{e} + \bar{\nu}_{er}) = \bar{\mathrm{n}} \rightarrow \bar{\mathrm{p}} + \mathrm{e}^+ + \nu_{el}$$

because P transforms a ν_{er} (which does not exist in nature) into a ν_{el} and thus amends the distortion.

Indeed, for some time physicists accepted the assumption that all interactions are invariant at least with respect to the combined operation CP. However, in 1964 it was shown by a group of physicists from Princeton University that even this principle does not always hold. Experimental proof for this was provided by the study of the decay of the neutral kaon. We have mentioned that there are two kinds of K^0. K^0_S which decays to give two pions, and the long-lived K^0_L which decays into three pions. It was found that out of about one thousand K^0_L particles, two on the average decay into two pions instead of three. A theoretical analysis of the phenomenon, much too lengthy to present here, showed that the invariance of CP is violated in such a decay process. The experiment was performed in the 33 GeV proton accelerator in Brookhaven and was conducted by J. W. Cronin and Val. L. Fitch. Both shared the 1980 Nobel prize in physics for their discovery. It is interesting to note that the experiment was conducted in order to provide hard experimental evidence for the invariance of CP. To the surprise of the physicists it proved the opposite – that certain interactions *are not* invariant with respect to CP. The physicists were bewildered by their own unexpected results. They kept their findings secret at first and checked and rechecked their data to make sure that their interpretation was not mistaken. But further experiments only served to confirm their conclusion. An immediate result of the fact that CP is not always conserved was that the invariance under T is also not absolute. Remember that the invariance under TCP always holds. This means that if an interaction is invariant under T, it must be invariant also under CP! And so, if CP is violated, this must be compensated by a violation of T. The situation, in the light of all these experiments is summarized at the end of Table 7.3.

Experiments such as this, which test an accepted theory to find out where it goes wrong (thus delimiting the regime of its validity) are very important. Karl Popper, the well-known philosopher of science, has claimed that such 'falsification' experiments make the progress of science.

It is possible that the T- or CP-violating processes are not ordinary weak interactions but represent the action of another force, the 'superweak interaction'.

We learn from the table that while weak interactions are not invariant at all with respect to C and P, the invariance under T (and thus also under CP) is only seldom violated.

Table 7.3. *Conservation laws and the basic interactions*

Physical quantity	Conserved in the interaction		
	Strong	Electro magnetic	Weak
Linear momentum	Yes	Yes	Yes
Energy–mass	Yes	Yes	Yes
Angular momentum	Yes	Yes	Yes
Electric charge	Yes	Yes	Yes
Baryon number, A	Yes	Yes	Yes
Electronic lepton number, L_e	Yes	Yes	Yes
Muonic lepton number, L_μ	Yes	Yes	Yes
Strangeness, S	Yes	Yes	No
I_3	Yes	Yes	No
Isospin, I	Yes	No	No
Parity, P	Yes	Yes	No
Charge conjugation, C	Yes	Yes	No
Time reversal, T (or CP)	Yes	Yes	Almost always yes
CPT	Yes	Yes	Yes

7.11 Conservation laws – summary

The conservation laws described in this chapter are summarized in Table 7.3. We can review the main points discussed so far by looking at that table.

The first four conservation laws, those of linear momentum, energy–mass, angular momentum and electric charge are laws of classical physics. They hold in particle physics as well, and govern all the four basic interactions. The remaining laws were unknown in classical physics. They were discovered in the course of experiments in the realm of elementary particles and their formulation involved the definition of new quantum numbers. The conservation laws of baryon and lepton numbers, strangeness and I_3 are simple 'counting rules' similar to the conservation law of electric charge. The isospin, on the other hand, is a vector quantity and its conservation law is somewhat more complex.

The baryon number and lepton numbers are conserved in all interactions observed so far, while the conservation of strangeness and I_3 pertain to electromagnetic and strong interactions, but not necessarily to weak interactions. The isospin conservation applies merely to strong interactions.

At the end of the table we find the three operations P, C and T. Applying any of these to the description of a strong or electromagnetic interaction always provides a process which can occur. However, when we apply any of them to a weak interaction we may

Table 7.4. *The relationship between basic interactions and particles*

	Basic interactions			
Type of particle	Strong	Electromagnetic	Weak	Gravitational
Hadrons	+	+	+	+
Electron and muon		+	+	+
Photon		+		+
Neutrinos			+	+
Graviton				+

+ means that the particle is affected by the interaction.

have a process which can never occur in reality. It can be shown, however, that the result of carrying out the three operations *P, C* and *T* successively (in any order) on a real life process is always a possible process.

From the invariance principle under *P*, a conservation law can be derived for the property named parity. Similarly, a quantum number (sometimes called the charge parity) can be defined which pertains to the behaviour of wave functions under the operation of the charge-conjugation (*C*). This quantum number is conserved in strong and electromagnetic interactions.

The conservation laws appearing in Table 7.3 suffice in explaining all the reactions between the particles we have met thus far. The discoveries of new particles in the 1970s, however, necessitated the definition of new 'charges' or quantum numbers and additional conservation laws, about which we shall learn in the next chapters.

Note the apparent gradation in the three basic interactions: the stronger the interaction, the greater the number of conservation laws which apply to it. Every conservation law which applies to a given interaction, also applies to all the interactions which are stronger than it. A similar gradation exists in the relationships between interactions and particles. The feebler an interaction, the greater the number of particles which are affected by it. If a certain particle is influenced by a given interaction, it is usually influenced by all weaker interactions as well (see Table 7.4). These two 'statistical' laws can be summed up as a single rule: The weaker an interaction, the more general it is, i.e. it acts on a greater number of particles and it is limited by fewer conservation laws. This interesting relationship is one of the indications that all four interactions might be just different levels of a single basic interaction. Much intensive study has been devoted towards attempts to unite the four basic interactions under one comprehensive basic theory, as we have already mentioned.

It should be emphasized again that the source of all conservation laws is experimentation. Their validity is continuously checked by carrying out new experiments and repeating more carefully the known old ones. This checking might reveal that a conservation law which was considered to be valid under all circumstances, is indeed violated in some processes. We described the discovery of the 'breakdown' of parity conservation in weak interactions, which caused a revolution in physical thinking. Attempts have been made to test other conservation laws such as the conservation of lepton numbers (by seeking the decay $\mu^- \to e\,\gamma$) and baryon number (by looking carefully for the decay of protons, see section 7.6). Up to now these attempts have been inconclusive.

Chapter Eight

Short-lived particles

We can divide the brief history of research on elementary particles into a number of periods, among them times of bewilderment and confusion and others of astounding discoveries; periods characterized by feverish study concentrated on a single subject, and periods during which progress was made on many fronts. During the years between 1947, the year the first V-tracks were discovered, and 1953, when the 'strangeness' idea was first proposed, experimental efforts were concentrated on the study of strange particles. Towards the end of the 1950s and during the early 1960s interest turned towards a new type of extremely short-lived particles, which were termed resonances.

8.1 Extremely short-lived particles

The reader who has gained some insight into the various conservation laws, and has understood how certain particles – such as strange particles – manage to live so long, thanks to conservation laws which prohibit decay by strong interactions, may now ask: Can't energetic collision (between protons for instance) also produce particles which are not barred by any conservation law from decaying by the strong interaction? If such particles do exist, their life expectancy is of the order of only 10^{-23} seconds (the time required for the effect of the strong force to traverse a distance equal to the diameter of the particle).

By the means available today there is no way to observe directly a particle with such a short life span. Even if its velocity approaches that of light, the distance traversed by such a particle until it decays is so short that we cannot distinguish between the point at which the particle is created and the point at which it decays.

You might suggest solving the problem of detecting such a particle by increasing its kinetic energy, thus expanding its lifetime through the relativistic effect of 'time dilation'. (It is thanks to this effect that the muons produced at high altitudes can reach the earth before decaying.) But a simple calculation shows that the energy required is much too high to be practical. For example, in order for a particle with a rest mass of 1 GeV and a lifetime of 10^{-22} seconds to be able to travel a distance of one-tenth of a millimetre before decaying, its kinetic energy should be of the order of 10^{10} GeV! The largest accelerators operating or planned today can provide the accelerated particles with only 10^3 GeV approximately. And even this is possible only after the particle has turned millions of times around the accelerator ring, the circumference of which is several kilometres – a mission impossible for a short-lived particle!

It is thus obvious that the track of a particle which decays strongly cannot be photographed in a bubble chamber or recorded by any other detector. Is there an alternative way of studying these particles? Should such short-lived beings actually be termed particles? The answer to these questions is that such extremely short-lived particles were in fact discovered, and despite their short lifetimes there is no reason to treat them as anything but respectable particles. In fact, amongst the particles we have already encountered there are several particles (π^0, η^0, Σ^0) which have shorter lifetimes than the others, as well as zero electric charges, which make their detection even more difficult. Nevertheless, these particles were identified and studied by investigating their decay products.

As we shall soon see, if a particle C is created through the reaction $A + B \rightarrow C$, and decays within 10^{-22}–10^{-23} seconds by the process: $C \rightarrow F + G$, the overall reaction observed will be: $A + B \rightarrow F + G$, but by probing this reaction under various conditions, the existence of particle C can be deduced, and its properties revealed.

8.2 Detecting methods

There are two main methods for detecting short-lived particles. In the first method, a beam of particles (such as pions) is fired at a target (a solid or liquid mass, for example) and the particles leaving the target are examined. A certain percentage of the incident particles will cross the barrier and will be emitted in their original direction, and with their original energy. Others will either be deflected in different directions, or leave with an energy different from the original. Some might not leave the target at all, having been replaced by new particles produced by the interaction between 'missile' and target. In each case we can assume that the incident particle was in contact with a nucleon in the target for a time span of the order of 10^{-23} seconds, which is the minimal time required for the interaction (or energy transfer) to take place. Can the two adjacent particles form a new particle during this brief period? Not necessarily. In order to define such a system as a particle, it must have certain physical properties: definite mass, definite angular momentum, and other definite quantum numbers. (The mass of such a system is not merely the sum of the rest masses of the colliding particles, since kinetic energy may also be transformed into mass at the moment of collision. Its spin is a function both of the spins of the colliding particles, and of their relative orbital angular momenta.)

Now imagine that we change the energy of the particles in the beam, and discover that at a given energy there is a sharp increase in the percentage of particles which are affected while passing through the target. This means that the two colliding particles – the one from the beam and the one in the target – now form a quasi-stable system, at a definite energy, or, in other words, when it possesses *a definite mass*. If we could show that it also possesses a definite angular momentum, definite parity and so on, we could seriously consider the possibility of granting it the status of a particle.

In 1952, Fermi, using an accelerator at the University of Chicago, examined the results of the scattering of pions from protons, and found that when the kinetic energy of the pion is about 195 MeV the probability of its traversing the target becomes particularly low. The results of the experiment are presented in Fig. 8.1. The 'rest mass' of the system $p\pi$ is represented by the horizontal axis; this rest mass is the sum of the rest masses of the two individual particles plus some of the kinetic energy possessed by the pion (it is quite easy to calculate – using the laws of momentum and energy conservation – how much of the energy turns into mass during the collision). Physicists call this quantity 'the centre-of-mass energy'. It can be seen that the probability of interaction between the pion and the proton is particularly high when the centre-of-mass energy equals 1230 MeV approximately. The results of this and other experiments showed that both π^+ and π^- can form with the proton a stable system with a mass of 1232 MeV.

During the 1950s, physicists were still hesitant to treat such states as real particles, and nicknamed them 'resonances'. The term resonance is commonly used in physics to describe the state in which a system can absorb energy with particularly great efficiency. To cite an example from acoustics: if you play a note in a room containing a piano, you will find that the piano strings begin to vibrate. The

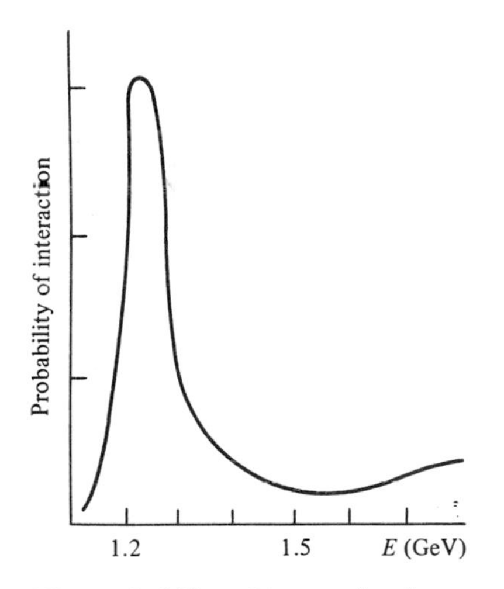

Figure 8.1. The probability of interaction between a beam of π^+ and stationary protons, as a function of the centre-of-mass energy (E). The prominent peak at the vicinity of 1.2 GeV is the resonance $\Delta^{++}(1232)$.

vibration of a string becomes particularly strong when its frequency is the same as that of the note played externally. We then say that the piano string 'resonates' with the external tone.

During the 1960s, as the number of known resonances increased and it was found that they do have not only definite masses, but definite spins, isospins and other quantum numbers, physicists began to treat them as particles in every sense of the word, and today the old term – resonance – is not widely used.

From scattering graphs such as the one shown in Fig. 8.1, one can determine the lifetime of the short-lived particle. Here we make use of the uncertainty principle which states $\Delta E \times \Delta t \approx h/2\pi$, where ΔE represents the uncertainty of the mass of the particle, and Δt its average lifetime (see section 2.5). For particles whose life spans are 10^{-10} seconds or more, ΔE is extremely small. But the shorter the time interval Δt, the greater the value of ΔE. In the scattering graph shown in Fig. 8.1 ΔE is simply the resonance width, i.e. the range of energies for which there is a large probability of generating the new particle. It is convenient to define ΔE as the width at half the height, as shown in Fig. 8.2. In Fig. 8.1 the resonance width is about 100 MeV, thus the lifetime equals

$$\Delta t \approx \frac{h/2\pi}{\Delta E} \approx \frac{6 \times 10^{-16}\ \text{eV} \cdot \text{second}}{10^{8}\ \text{eV}} \approx 0.6 \times 10^{-23}\ \text{second}$$

and this is a typical lifetime of a particle decaying via a strong interaction.

$$\Delta t \sim \hbar/\Delta E$$

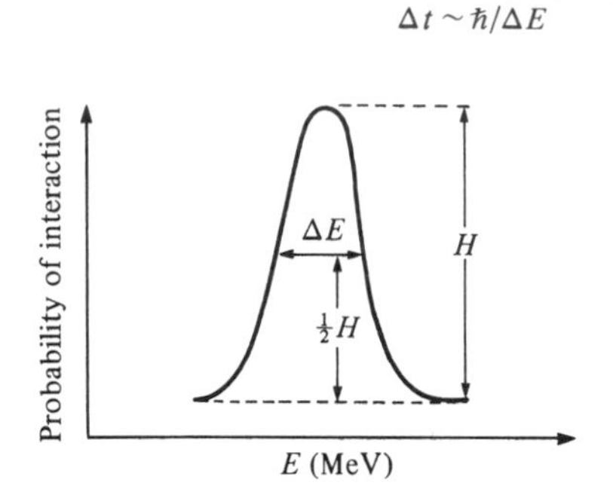

Figure 8.2. From the width of the resonance at half of its height, the lifetime of the particle can be deduced.

We will now describe, with the help of an example, the other method of studying short-lived particles. Suppose that you send a beam of π^+ of a certain energy into a cloud chamber, and obtain many photographs of the reaction $\pi^+ + p \rightarrow p + \pi^+ + \pi^0$. Let's assume that you have reason to suspect that the process actually takes place in two stages:

$$\pi^+ + p \rightarrow p + \rho^+$$
$$\rho^+ \rightarrow \pi^+ + \pi^0$$

where ρ^+ is an extremely short-lived particle (so much so that its track cannot be seen in the photographs). How can you check your hypothesis?

By measuring how much the tracks are deflected by a magnetic field, the energy and momentum of the reaction products can be calculated. (The track of π^0 is not visible in the photographs, but its energy equals the difference between the original energy and the sum of the energies of the other products. Moreover, in a big bubble chamber the tracks of the two electron–positron pairs, which are the final products of the π^0, can be recorded.) Now one can easily determine the rest mass of the hypothetical particle which *may have* decayed into the two pions (its mass would equal the sum of the masses of the pions plus a known part of their kinetic energies). We now carry out similar calculations for many such photographs of this type of experiment, and compare our results. If the values obtained in all the cases are similar, it does not stand to reason that the energy is similarly divided among the reaction products as a matter of coincidence, just as it is not reasonable that a roulette wheel (which has not been 'fixed') comes up with the same number tens and hundreds of times. We look at such a situation as evidence that the presumed short-lived particle, named ρ^+, does indeed exist.

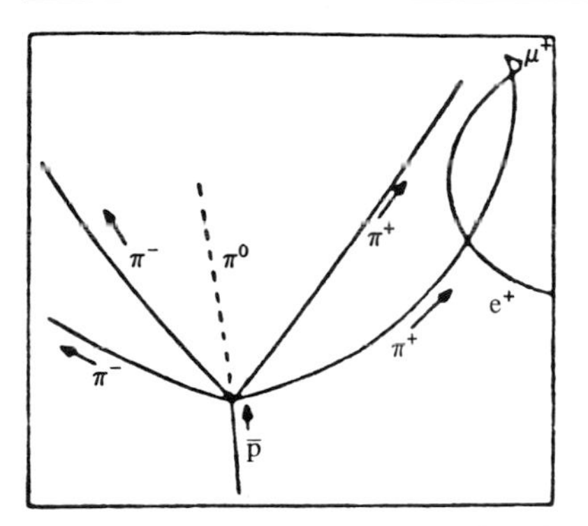

Figure 8.3. A reconstruction of a bubble chamber photograph of a $p\bar{p}$ annihilation producing five pions.

More intricate processes can be analysed in the same manner. Fig. 8.3 reconstructs a bubble chamber photograph of the reaction:

$$\bar{p} + p \rightarrow 2\pi^- + 2\pi^+ + \pi^0$$

800 photographs of this reaction were examined in 1960 by L. Alvarez's group at Berkeley. The aim was to test the assumption that three of the pions were the decay products of a neutral short-lived particle. The problem was that one could not pick the three decay products out of the five recorded pions. Therefore, for any photographed reaction, the energies of all the possible combinations of three pions with a total charge of zero were calculated on a computer. Fig. 8.4 depicts the energy distribution of the possible π^+, π^-, π^0 trio of pions. The horizontal axis is the energy axis and the vertical axis represents the number of trios possessing a certain

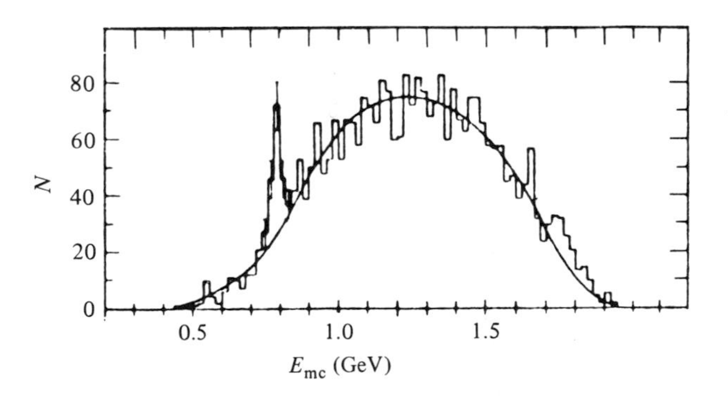

Figure 8.4. Energy distribution of trios of pions in 800 bubble chamber photographs of the reaction depicted in Fig. 8.3. E_{mc} is the centre-of-mass energy and N is the number of events. (Maglič, Alvarez, Rosenfeld & Stevenson (1961), *Phys. Rev. Lett.* vol. 7, 178.) At 0.77 GeV we observe the ω^0 peak.

value of energy (centre-of-mass energy). Were all the five pions created in the annihilation of the proton and anti-proton, the distribution of energy between them should be random. Theoretical calculations show that in this case the graph would display small fluctuations around a bell-shaped smooth curve. As shown in Fig. 8.4, in most of the energy range the experimental results really oscillated around such a curve, but from the resonance or peak in the vicinity of 0.8 GeV one can learn that at least in some of the cases the reaction that occurred was actually

$$\bar{p} + p \rightarrow \omega + \pi^+ + \pi^-$$
$$\omega \rightarrow \pi^+ + \pi^- + \pi^0$$

where ω (omega) denotes a neutral short-lived meson. The resonance is even more prominent in Fig. 8.5, which shows the difference between the experimental results and the theoretical smooth bell curve. In this and similar experiments the existence of the ω was proved beyond doubt, and its properties studied in detail. Its mass

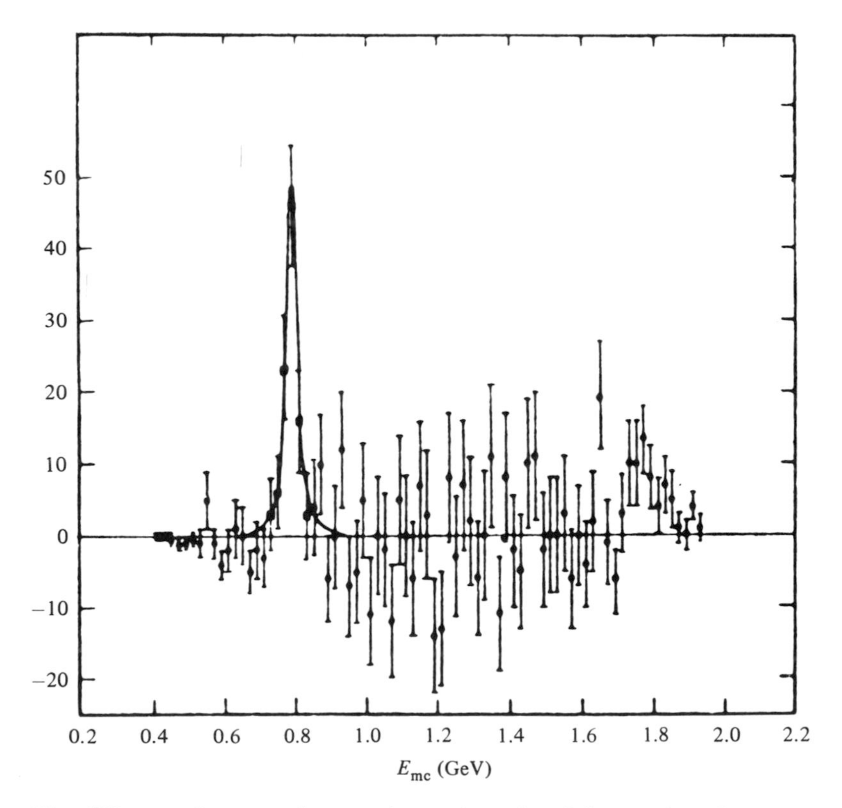

Figure 8.5. The difference between the experimental results of the previous figure, and the smooth bell-shaped curve. The ω^0 peak is very prominent.

was found to be 787 MeV and its lifetime 7×10^{-23} seconds. Despite the fact that the track of the ω was not recorded in any particle detector, we are sure that if we could take photographs such as the one described in Fig. 8.3, but magnified 10^{12} times, we would find the reaction of the type shown in Fig. 8.6.

One can explain this detection method by the following analogy: many children are rushing about in the schoolyard with candies in their hands. You know that each pair of children shared a packet of candies, not necessarily in equal proportions. The openhanded ones gave their companion most of the packet. Other, less generous ones tried to get most of the candies and make their companion be satisfied with one or two pieces. In many cases however, the division was more or less fair. In order to find out how many candies each packet contained, you can 'interview' as many children as possible and write down the number of candies held by each. Then you calculate the total number of candies possessed by all the possible pairs, and draw a graph of the number of pairs as a function of the number of candies of each pair. If the division was totally random, you would get a bell-shaped curve. But since a proportion of the pairs you combine are 'genuine', and the sum of candies of such a pair equals the number of candies in the original packet, a peak or 'tower' will usually appear on the bell curve at that value (i.e. the number of candies in an original packet).

8.3 More and more resonances

The first resonance discovered by Fermi (Fig. 8.1) is a member of a group of four particles denoted Δ (delta) which will be described later. This discovery was like the first swallow which does not yet bring tidings of the coming spring. In the 1950s the research of short-lived particles was carried out slowly, and its contribution to the list of known particles was meagre. In the early 1960s, however, the study of resonances began to bear fruits and during the 1960s and 1970s hundreds of short-lived particles were added to the

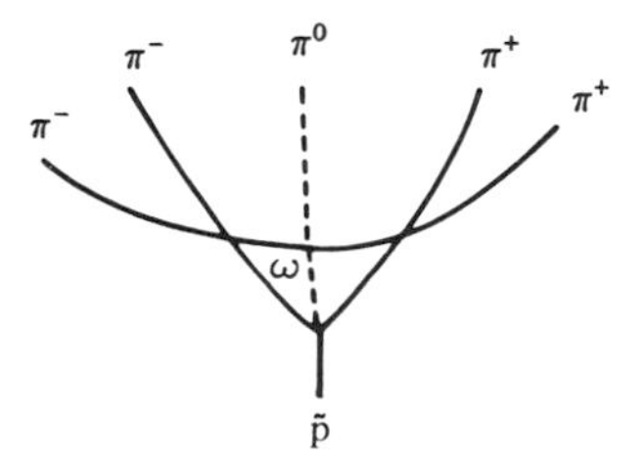

Figure 8.6. The picture we would get if we could photograph the reaction of Fig. 8.3, magnified 10^{12} times.

ever-growing list. The inauguration of each new accelerator producing proton or electron beams at energy ranges unattainable before, as well as the use of large and more effective bubble chambers, was almost immediately followed by the discovery of new resonances. When the Nobel prize was granted in 1968 to Luis Alvarez of Berkeley, his group's dominant role in the discovery of new resonances was highly commended by the Nobel referee (this was in addition to his part in the development of the bubble chamber, together with D. Glaser who received the prize in 1960).

A remarkable fact is that almost all the new particles were hadrons – particles participating in the strong interaction. New leptons were not added to the list of elementary particles until the end of the 1970s (a new heavy lepton was then discovered, a story we relate in Chapter 10).

The list of new particles included both baryons and mesons. All the resonances discovered up until 1974 followed the rules applied to the 'semi-stable' hadrons appearing in Tables 5.1 and 5.2 and the classification described in Chapter 6. Their production and decay processes were found to be governed by the same additive ('counting') conservation laws we described in Chapter 7, including the equations linking the various quantum numbers. The hundreds of new particles which were added to the list represented multiplets similar to the ones with longer lifetimes which are listed in Tables 5.1 and 5.2.

The most significant point was the close similarity usually found between each resonance and a 'semi-stable' particle (see the table of semi-stable particles on p. 264). For example, a resonance at 1915 MeV, was found to have a strangeness of -1, baryon number $+1$, spin $\frac{5}{2}$ and parity $+1$, and to appear in three states of charge: $+1$, 0 and -1 (or an isospin of 1). Except for mass and spin, this particle is identical in all respects to the Σ. Actually it may be considered to be an 'excited state' of the Σ. This reminds us of a similar situation in atomic and nuclear physics. A sodium atom which absorbs energy and re-emits it in the form of a photon 10^{-8} seconds later does not turn into a new particle during this interval, but is just an excited sodium atom. This is also the case with a nucleus which absorbs energy and emits it after a short time in the form of a gamma photon. The excited nucleus has the same mass number (A) and atomic number (Z) as the unexcited nucleus, but its angular momentum and its mass are larger. To show that the resonance at 1915 MeV can be considered an excited state of the Σ, it is denoted by $\Sigma(1915)$.

Similar relationships could be found between most of the resonances discovered up until 1974, and baryons or mesons in the table of semi-stable particles on p. 264. At least a dozen groups of resonances denoted by Σ have been found in energies up to 3000 MeV. All of them have the same strangeness, baryon number and isospin as the semi-stable Σ, and could be looked upon as excited Σs. Similar excited states were discovered for other baryons as well. For example, a group of resonances was found to have an isospin of $\frac{1}{2}$ like the nucleons (they appeared in pairs of N^+, N^0) and were identified as excited nucleons. The Δ mentioned above (Fermi's resonance) was found to form a multiplet of four particles (isospin $\frac{3}{2}$), one of which possessed a *double charge*: Δ^{++}, Δ^+, Δ^0, Δ^-. Other such Δ multiplets were found with higher masses. The lowest state of this group, Δ(1232), cannot be regarded as an excited state of a group in the list of 'semi-stable' baryons. The Δ(1232) itself is a resonance with a lifetime of about 10^{-23} seconds, and is the ground state of the heavier and higher-spin Δ multiplets.

Mesonic resonances were discovered as well. The short-lived η^0 (3×10^{-19} seconds) is considered to be a 'semi-stable' particle even though its research was carried out using methods of resonance study. The reason for this is that it decays via the electromagnetic interaction, and not via the strong interaction, and thus exists at least 1000 times longer than a common resonance.

The lightest mesonic resonance is a trio of particles of spin 1, near 770 MeV, which are known as ρ^+, ρ^0, ρ^-. Excited states of the pions were discovered at higher energies. There are resonances similar to the kaons, the first of which is at 892 MeV, and resonances similar to the η^0, among them $\eta'(958)$ of spin 0, ω(787) and φ(1020) of spin 1, $f^0(1270)$ and $f'(1515)$ of spin 2, and so on. (The number in parentheses always indicates the mass in MeV.) These are only a few examples of the hundreds of baryonic and mesonic resonances which have been discovered so far.

A theory published by the Italian physicist Tullio Regge in 1959, and perfected by the Americans G. Chew and S. Frautschi in 1961, assisted in explaining most of these resonances as excited states of the lower-energy systems. This theory links the masses of the resonances and their spins, and predicts that every hadron should have excited states at higher energies, at intervals of two additional units of angular momentum. Many such examples were actually found experimentally. More light is shed on the relation between resonances and semi-stable particles by the theory of quarks discussed in Chapter 10.

8.4 Storage rings

To conclude this chapter we shall describe an important development in the technology of accelerators which significantly influenced the research of the short-lived particles. In several accelerator centres throughout the world, a new type of machine was designed and constructed during the 1960s in order to study what happens when two high-energy particles moving in opposite directions meet in a head-on collision. In these facilities, known as 'storage rings' or 'colliding-beam accelerators' or just 'colliders', some outstanding discoveries were made. To understand the importance of this new idea, a short introduction is needed. (See also section 4.2.)

Think of a fast-moving particle, A, interacting with a stationary particle, B, so that both disappear and a new particle C is created. Owing to the conservation of linear momentum, the velocity of C cannot be zero – it must carry all the momentum previously possessed by A. Particle C is thus produced with a certain kinetic energy. Consequently, not all the kinetic energy of A can turn into *the mass* of the new particle; a substantial part of it remains in the form of kinetic energy.

If, for example, a proton with a velocity much less than that of light hits a stationary proton, only about half the kinetic energy can be converted into the mass of new particles. With relativistic velocities the situation is even worse. If a proton (the rest mass of which is close to 1 GeV) is accelerated to a total energy of 200 GeV, and then collides with a proton at rest, only about one-tenth of its energy can turn into the mass of new particles. The rest of the energy invested in the acceleration of the proton is a sheer loss. Practically, in order to produce a resonance at 20 GeV in such a collision, a 200-GeV accelerator is needed. (The amount of energy that can be converted into mass is referred to as 'the centre-of-mass energy'.)

The situation is quite different when two particles possessing equal masses and travelling in opposite directions with the same speed meet in a head-on collision. In this case the total momentum is originally *zero* and the conservation of momentum does not limit the transformation of kinetic energy into mass. Since the new particles produced in the reaction can be created *at rest*, actually all the kinetic energy of the colliding particles can turn into mass, to make new products. The centre-of-mass energy is simply the total energy of the particles. A 10-GeV accelerator is therefore adequate to produce a short-lived particle at 20 GeV, provided that we can arrange head-on collisions among the accelerated particles. Such a collision would be especially effective if it occurs between a particle and its corresponding anti-particle; for example – between an electron and a positron, or a proton and an anti-proton. The reaction which takes place is an annihilation, in which all the mass of the two particles is converted into energy. In this process all the original quantum numbers are abolished (because the quantum numbers of the two participants are opposite in sign) and the total energy can materialize into new particles entirely different from the original ones.

The first electron–positron storage rings were inaugurated during the years 1963–67 at Frascati (Italy), at Stanford University in California, at Novosibirsk in the USSR, at Orsay near Paris, and at Cambridge, Massachusetts, in the USA. The largest machine, in Cambridge, reached a total energy of 7 GeV (3.5 GeV in each beam). In most of those machines the electrons and positrons circulated in the same ring in opposite directions, but tangent or interlaced rings were also employed.

Following the success of the first storage rings, larger 'second generation' machines were built in the early 1970s. They included SPEAR (the Stanford Positron Electron Accelerating Ring) at the Stanford Linear Accelerator Center (SLAC), and Doris at the DESY (Deutches Elektronen Synchrotron) laboratory in Hamburg. Both reached a total energy of about 9 GeV. These devices were distinguished not only by higher energies than the storage rings of the previous generation, but by better energy resolution and more sensitive detectors as well. The energy range they spanned happened to be particularly rich in new important phenomena as we shall see in Chapter 10.

A 'third generation' of storage rings which appeared towards the end of the 1970s included PETRA in Hamburg (total energy of 38 GeV, later increased to 46 GeV), CESR (Cornell Electron Synchrotron Ring) at Cornell University, USA (16 GeV), and PEP at SLAC (36 GeV).

The next generation of electron–positron colliders include the conventional storage ring structure, as well as innovative schemes. At CERN the 27-kilometre circumference LEP ring is planned to be operative by 1989 with a total energy of 100 GeV, later to be pushed up to 250 GeV. (For comparison, the circumference of SPEAR is only 0.25 kilometres and that of PEP is about 2 kilometres). A similar ring is to be constructed at the KEK laboratory in Japan (the TRISTAN collider, with a total energy of 60 GeV). A new idea is going to be tested at SLAC. The existing linac will serve to accelerate beams of electrons and positrons which will be deflected in opposite directions and then directed towards each other and collide head on at a total energy of 100 GeV (such a device needs a very fine tuning in order to obtain focused beams in a single pass). At Novosibirsk in the USSR an ambitious project has been designed: to use two opposing linacs in order to fire electrons and positrons at each other at a total energy of 300 GeV.

Proton–anti-proton storage rings began operating only in the early 1980s. The SPS (Super Proton Synchrotron) at CERN was modified into a $p\bar{p}$ storage ring with an impressive total energy of 540 GeV. Beginning operating in 1981, this machine soon fulfilled the hopes of its designers and in 1982–3 provided the first evidence of the weak force carriers (see section 10.8). At Fermilab near Chicago, newly installed superconducting magnets turned the old synchrotron into a proton accelerator of 1000 GeV (=1 TeV, or *tera* eV). This machine is named the Tevatron and is to be converted (by 1986) to a 2000-GeV $p\bar{p}$ storage ring. An even bigger $p\bar{p}$ collider is planned in the USSR: a 3000-GeV proton accelerator to be built at Serpukhov near Moscow by 1990 will later be turned into a 6000-GeV $p\bar{p}$ storage ring.

Most but not all of the colliders employ collisions between particles and their corresponding anti-particles. Interesting physical results have been produced since 1971 at the intersecting storage ring pp machine at CERN, where interlaced rings are used to make protons collide with protons at total energies up to 62 GeV. At DESY (Hamburg), a machine named HERA is being constructed in which 30-GeV electrons will collide head on with 820-GeV protons.

In the next section the structure of SPEAR at Stanford is described. This electron–positron storage ring was the scene of several important discoveries in the 1970s, and may best represent this type of machine. Further reference to the operation of storage rings will be made in Chapter 10.

The structure of SPEAR

The central part of SPEAR is a vacuum aluminium tube with a cross-section of several square centimetres. The tube forms an elliptical ring with a mean diameter of 80 metres (see Fig. 8.7). When the machine is operating, two bunches of particles – one of electrons and the other one of positrons – are moving in the tube in opposite directions. Each bunch is a few centimetres long and less than a millimetre thick and it contains some 10^{11} particles. Both bunches travel at a speed close to the speed of light and they pass through each other twice in each cycle (or two million times in a second). These encounters occur at two opposite sections of the ring – the interaction regions – where the tube is surrounded by detectors

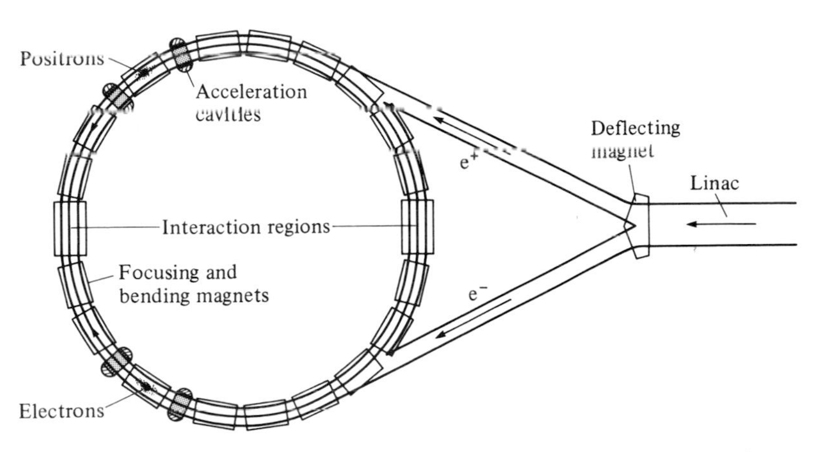

Figure 8.7. A schematic description (not to scale) of the SPEAR storage ring at the Stanford Linear Accelerator Center (SLAC).

which record the products of the annihilations. In other places along the ring, electromagnets of two types are installed. The first deflects the tracks of the particles into the required elliptical orbit and the other focuses the beams and keeps them narrow. At four points along the ring, radio-frequency electromagnetic fields are used to accelerate the particles and provide them with the energy they lose by emitting electromagnetic radiation.

The electrons and positrons are supplied by the SLAC 3-kilometre linear accelerator. The electrons are transferred into the storage ring directly from the linear accelerator. In order to produce the positrons, electrons are made to hit a copper target at the end of the first third of the accelerator. The result is a shower of photons, part of which later produce electron–positron pairs. The electric fields in the continuation of the accelerator are adjusted to accelerate only the positrons, and drive them into the storage ring. The whole process of filling the ring with the required amounts of electrons and positrons takes only a few minutes, after which they may go around the ring for several hours. The energy of the particles can be adjusted to values between 1.2 and 4.5 GeV, which gives a total energy of between 2.4 and 9 GeV. The particles investigated in the storage ring are short-lived resonances, and only their decay products reach the detectors.

The two methods for detecting resonances mentioned above can be used. One can gradually alter the energy of the beams and check whether there is a considerable enhancement in the annihilation rate at some specific energy. It should be mentioned that although the two beams, each containing 10^{11} particles, cross each other two million times a second, the probability of an interaction between an electron and a positron is quite small and the usual rate of annihilations is one every few seconds. If this rate increases at a certain energy, it is noticed at once. When the detectors begin to tick every single second or less, it is clear that a new short-lived particle is produced at this energy.

The alternative method for detecting resonances can also be easily employed in storage rings. An array of detectors at SPEAR ('the magnetic detector', see section 10.2) which was used for some important discoveries is capable of measuring the energy of the annihilation products to great precision. One can examine the energy distribution of a certain combination of products (e.g. $\pi^+\pi^-$, or πK, etc.) and find out whether these energies are concentrated around some specific values and stand out above the smooth distribution curve. Performing such an experiment in a storage ring might require a lot of patience, since the collection of data – especially when rare events are concerned – may last months or years.

Further developments in colliding-beam accelerators and the exciting discoveries made in these machines are described in Chapter 10. The next chapter discusses some important theoretical advances made in the 1960s which had a profound influence on the world picture of particle physics.

Chapter Nine

To the quarks – via the eightfold way

In the 1960s and 1970s the list of hadronic resonances rapidly expanded. Physicists watched its growth with mixed feelings. On the one hand the many new particles would provide them with a rich field of investigation for many years to come, but on the other, the inability to explain the connections between the various particles pointed only too clearly to the lack of a comprehensive theory concerning the nature of sub-atomic particles. The classification into four families no longer provided a means of bringing order into the chaos, since two of the families – the mesons and baryons – had flourished, and eventually added tens upon tens of new members to their number. In the late 1950s and early 1960s several attempts were made to draw up a new classification scheme which would divide the two large hadronic families into sub-families on the basis of the quantum numbers of the particles. Such a scheme was needed not only as an aid to finding one's way around the jungle of particles, but it could also be the first step towards finding a satisfactory theory which would explain the existence and properties of all the particles.

Experience had shown that when certain observations could not be fitted into a comprehensive theory, it was often useful to sum up the abstracted regularities in heuristic formulae or illustrative tables, even if their theoretical basis was not yet clear. Physicists speak in this case of 'empirical laws' or 'phenomenological theory'. This approach has often led ultimately to the discovery of the fundamental law. A case in point relates to the wavelengths of light emitted by gases when excited by heat or electric current. The question which had intrigued the physicists of the nineteenth century was why a certain gas emitted particular wavelengths and not others. Attempts were made to develop formulae which would interrelate the various wavelengths characterizing certain gaseous substances. Balmer's formula for lines in the hydrogen spectrum was particularly simple, and its analysis led eventually to the Bohr model of the atom, which provided, *inter alia*, the principle explaining the relation between the structure of the atom and its spectrum. In a similar way, Mendeleev's periodic table led to the elucidation of the atomic structure of the elements, and Planck's heuristic formula for the spectral composition of black body radiation brought about the discovery of the light quantum.

9.1 The Sakata model

One of the important attempts at bringing order into the world of hadrons was made by the Japanese physicist, S. Sakata, mentioned above (section 3.5) for his part in solving 'the two mesons enigma'. In the 1950s and 1960s Sakata and M. Taketani gathered around themselves a group of theoretical Japanese physicists who made an important contribution to particle physics.

In 1956 Sakata proposed a model according to which all hadrons were composed of combinations of six known particles: the neutron (n), proton (p) and lambda (Λ^0) and their corresponding anti-particles. Table 9.1 shows how the model explained the semi-stable mesons discovered by that time as built of pairs of particle and anti-particle from among those six building blocks.

Table 9.1. *The mesons according to the Sakata model*

	$\bar{p}$ $A=-1$ $Q=-1$ $I_3=-\frac{1}{2}$	$\bar{n}$ $A=-1$ $Q=0$ $I_3=\frac{1}{2}$	$\bar{\Lambda}$ $A=-1$ $Q=0$ $I_3=0$
p $A=1$ $Q=1$ $I_3=\frac{1}{2}$	π^0, η^0 $A=0$ $Q=0$ $I_3=0$	π^+ $A=0$ $Q=1$ $I_3=1$	K^+ $A=0$ $Q=1$ $I_3=\frac{1}{2}$
n $A=1$ $Q=0$ $I_3=-\frac{1}{2}$	π^- $A=0$ $Q=-1$ $I_3=-1$	π^0, η^0 $A=0$ $Q=0$ $I_3=0$	K^0 $A=0$ $Q=0$ $I_3=\frac{1}{2}$
Λ^0 $A=1$ $Q=0$ $I_3=0$	K^- $A=0$ $Q=-1$ $I_3=-\frac{1}{2}$	$\bar{K}^0$ $A=0$ $Q=0$ $I_3=\frac{1}{2}$	η^0 $A=0$ $Q=0$ $I_3=0$

A = baryon number.
Q = electric charge.
I_3 = z-component of the isospin.

Sakata's model actually generalized an idea suggested by E. Fermi and C. N. Yang in 1949, of regarding the pions as composed of nucleons and their anti-particles.

According to this model π^- is a bound state of n and $\bar{p}$, π^+ is composed of p and $\bar{n}$, and so on. The spins in each system are opposite in direction and cancel each other out (as we know, all the mesons in the table have zero spin). The table shows some of the quantum numbers of the particles: the baryon number (A), the electric charge (Q) and the z-component of the isospin (I_3). Each of these quantum numbers of the composite particle equals the sum of the corresponding quantum numbers of the constituents. Similarly, it is possible to 'build' the known baryons from triplets of the particles n, p, Λ^0, $\bar{n}$, $\bar{p}$, $\bar{\Lambda}^0$. In fact, Fermi and Yang had investigated the possible binding mechanism back in 1949.

The masses of certain particles, according to this model, are much smaller than the sum of masses of their building blocks. For example, the masses of a proton and anti-proton add up to 1876.6 MeV, whereas the mass of the π^0 formed from them, according to Sakata, is a mere 135 MeV. This should not be considered a flaw in the model. We must remember that the law according to which the mass of a system always equals the sum of the masses of its components is a law of classical physics. This law lost its validity with the advent of special relativity, which allows conversion of mass into energy and vice versa. Thus in nuclear physics one is familiar with the fact that the mass of the nucleus is less than the sum of the mass of its nucleons, the difference being the so-called 'binding energy'. The nuclear binding energy, however, forms only a small fraction of the mass of the nucleus, while in Sakata's model the binding energy would be several times greater than the mass of the composite particle. Nevertheless, this is not in itself a contradiction of any law of physics.

When a 100-tonne spaceship is launched it has to overcome the earth's (gravitational) binding energy. The earth–spaceship system before the launch has a smaller mass than the summed masses of the earth and the free spaceship, but this difference is only about 0.1 grams, or 10^{-34} times the total mass!

Sakata's model received a mixed reception among physicists. Some associated it with the kind of breakthrough that occurred in atomic physics when it was found that all atoms were built of neutrons, protons and electrons. Others were more sceptical, and recalled Prout's theory of the early nineteenth century, which postulated that all the elements were built of hydrogen atoms. This was a bold idea, and close to the truth, but was incomplete because the neutron was unknown at that time. Had the theory been accepted it might have set back the progress of chemistry, by strengthening the alchemists' hope of converting elements into one another by chemical processes. Prout's theory was, however, rejected because several of the elements then known had non-integral atomic weights – notably 35.5 for chlorine, 107.8 for silver, and so on. This was taken as conclusive evidence that they could not be composed of hydrogen atoms. (Today we know that this reasoning was false. The non-integral atomic weights stem from the co-existence of different isotopes of the same element.) One can regard Prout's idea as an example of an over-hasty jump from a phenomenological theory to a structural model, i.e. a model which explains the properties of a system in terms of its structure.

9.2 The eightfold way

In 1961, a new model for classifying the hadrons was proposed independently by Murray Gell-Mann from Caltech (California Institute of Technology) and by Yuval Ne'eman, one of the authors of this book, who was working in Britain at the time. This model discarded the idea that some of the hadrons were the building blocks of all the rest. Instead it was based on the division of the hadrons into 'families' or supermultiplets, and the finding of connections between the various members of each family, using the mathematical notion of groups. For reasons which will be explained later the model was named 'the eightfold way'.

In Chapter 7 we saw that the classification of the hadrons into multiplets, each of which contained similar particles differing in electric charge, proved successful. This division made possible the definition of two 'good' quantum numbers (i.e. quantum numbers which were found to be conserved in certain interactions): the isospin I, and its component I_3. The eightfold way was based on uniting a number of multiplets – those having the same baryon number, spin and parity – into a larger set or supermultiplet.

The family of mesons with zero spin in Table 5.1 is an example of such a supermultiplet. It consists of an octet containing four multiplets – a triplet of pions, a doublet of kaons and its anti-particle conjugate doublet (with inverted quantum numbers) and the single eta meson. All have a baryon number of zero (like all mesons), zero spin and negative parity. One could say that if the division of hadrons into multiplets revealed that the pions were a family of three brothers and the kaons a family of two brothers and their 'anti-brothers', the eightfold way regarded the pions, kaons and eta meson as first cousins.

The grouping of the hadrons into supermultiplets is only the first step. The second step is to represent each particle in the supermultiplet by a point on a coordinate system where one axis represents the quantum number I_3, and the other axis the strangeness S. In such a coordinate system our pion–kaon–eta octet yields a symmetrical hexagonal pattern, with one particle at each vertex and two particles at the centre (see Fig. 9.1). Note the symmetry of the

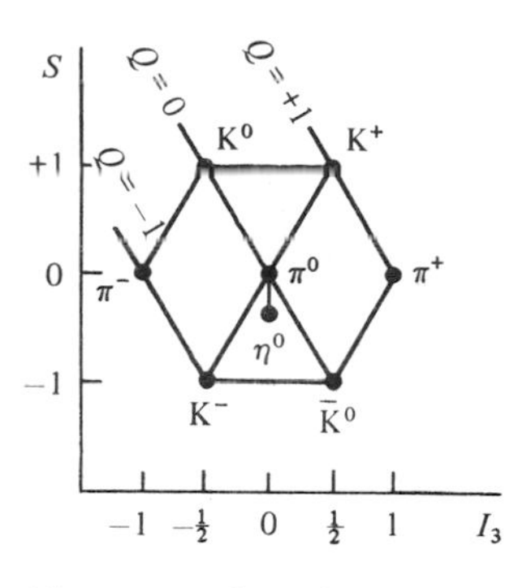

Figure 9.1. The octet of semi-stable mesons (spin 0).

electric charge resulting automatically from this arrangement: the four neutral particles of the octet lie along the diagonal, with the positively charged particles on one side and the negatively charged particles on the other.

This rather beautiful geometric shape is quite surprising, but even more startling is the fact that the semi-stable baryons of spin $\frac{1}{2}$ of Table 5.2 (excluding the Ω^-, which has spin $\frac{3}{2}$) yield an identical pattern in the I_3–S plane. This group consists of eight particles, all possessing the baryon number +1, spin $\frac{1}{2}$ and positive parity. The pattern they form is shown in Fig. 9.2. The eight anti-particles of this group make a similar pattern. (In the meson octet, the particles and their anti-particles appear together, since the baryon number – which is one of the criteria for classification – is zero for both mesons and anti-mesons.)

Mesonic and baryonic resonances which were discovered in the 1960s fitted well into additional supermultiplets, most of them octets, but some of them decuplets and some singlets. All form

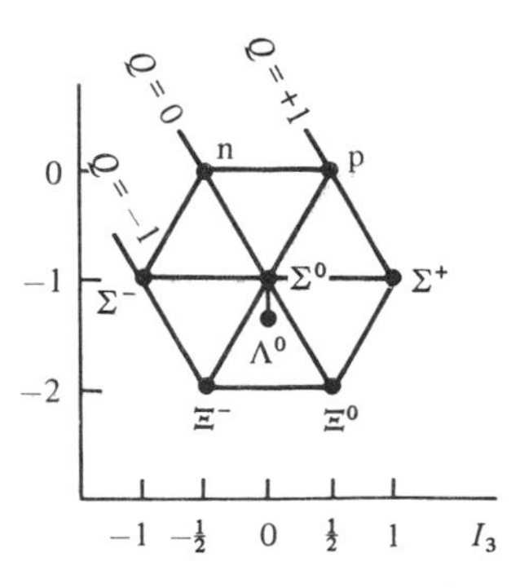

Figure 9.2. The octet of semi-stable baryons (spin $\frac{1}{2}$).

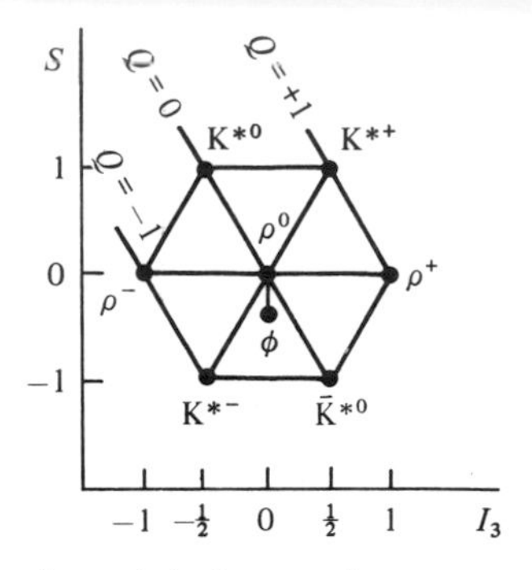

Figure 9.3. An octet of mesonic resonances (spin 1). K* denotes a group of resonances similar to the kaons.

symmetrical patterns in the I_3–S plane. Fig. 9.3 describes a group of mesonic resonances with spin 1, and Fig. 9.4 a decuplet of heavy baryons (this is the famous decuplet of the Ω^- which will be discussed later). The η', ω and φ are examples of singlets. In fact, all the hadrons discovered up until 1974 fitted well into this scheme, and formed similar symmetrical patterns in the I_3–S plane.

Was this just a coincidence that the hadrons formed such symmetrical shapes, or did it reflect something deeper, some fundamental structure of nature which manifested itself through the geometric patterns? Whatever the answer to that question might turn out to be, the inventors of the eightfold way showed that the discovered symmetry enabled the definition of a new quantum number – unitary spin – which was conserved under strong interactions. We shall not go into the definition of unitary spin here, except to point out that it is a system of eight components, each of which is a combination of quantum numbers we have already met. On the strength of these eight components, the new model was dubbed 'the eightfold way', by association with 'the noble eightfold way' of

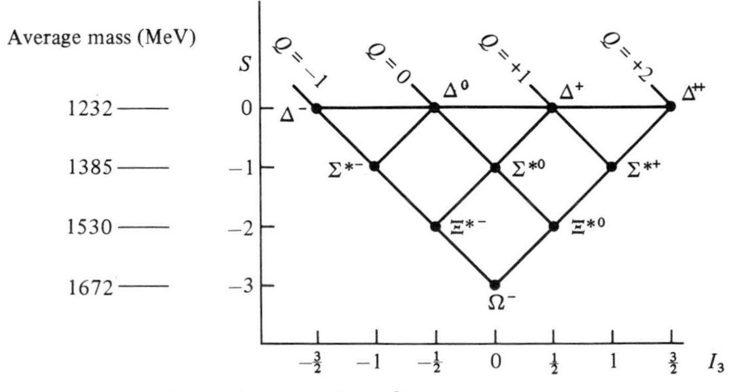

Figure 9.4. A decuplet of heavy baryons (spin $\frac{3}{2}$).

Buddhism (the way a person should walk through life in compliance with the 'eight commandments' of the Buddhist religion).

Gell-Mann and Ne'eman showed that, just as the particles of a particular multiplet could be regarded as different states of the same basic particle, differing only in the 'direction' of the isotopic spin, so the particles populating the supermultiplet could be seen as different states characterized by different 'directions' of the unitary spin. 'Rotation' of the unitary spin in the imaginary space in which it is defined changes one particle into another.

The mathematical formulation of the model drew on a (then) little-known branch of the theory of groups, a mathematical theory whose previous applications in physics had dealt with the symmetries of crystals. This special branch had been developed in the nineteenth century by the Norwegian mathematician Sophus Lie. The 'Lie groups' which for about 100 years had found no practical application, became the cornerstone of the new physical theory. The specific 'group' used here was known by the mathematicians as 'the group of three-dimensional special (unimodular) unitary matrices', abbreviated to $SU(3)$, and this name has become a synonym for the model of the eightfold way.

The formalism of $SU(3)$, using the diagrams in Figs. 9.1–9.4, imposes precise relationships between various properties of the particles in a supermultiplet (such as masses and magnetic moments). Thus, for example, the following connections result between the masses of the semi-stable baryons:

$$m(\mathrm{n}) + m(\Xi^0) = \tfrac{1}{2}\{3\,m(\Lambda^0) + m(\Sigma^0)\}$$
$$m(\Xi^-) - m(\Xi^0) = m(\Sigma^-) - m(\Sigma^+) + m(\mathrm{p}) - m(\mathrm{n})$$

These two equations – known respectively as the Gell-Mann–Okubo and the Coleman–Glashow equations – are satisfied to a great numerical accuracy, as the reader can verify for himself.

Another conclusion that can be drawn from the model is that the strong force originates in 'charges' like the electric force, except that instead of two charges and one carrying particle (the photon), we have here a more complicated system of several 'charges' and eight carrying particles. The new approach to the strong force made it possible to calculate coupling constants (which determine the intensity of the interaction between two particles) and probabilities of the occurrence of reactions.

A group is a collection of transformations or operations which fulfil certain conditions. For example, a combination of two operations must yield an operation which is itself included in the group. Rotations in three-dimensional space, for example, form a group. We can take an object, such as a book or plate, and rotate it in any direction; a combination of two rotations performed consecutively is also a rotation, which could have been performed in one operation.

In 1951, Yoel (Giulio) Racah of the Hebrew University in Jerusalem visited the Princeton Institute for Advanced Studies, where he presented a series of lectures on group theory; he himself had used the theory to solve important problems in atomic spectroscopy. Among the audience were Gell-Mann, Salam, Abraham Pais, and also Lee and Yang (who later demonstrated the non-conservation of parity). Racah showed how to classify the Lie groups and pointed out that they could be divided into four families, of which three represented rotations in space of real dimensions,

9.3 Discovery of the omega minus

At the time that the eightfold way was proposed, not all the particles in Figs. 9.1 to 9.4 had been discovered, and in fact not one of the while the fourth described generalized rotations in space whose dimensions were complex numbers (i.e. containing the imaginary number $\sqrt{-1}$).

Unfortunately this fourth family was overlooked when some years later attempts were being made to use groups in order to classify the hadrons. Only rotations in real space were examined, and these failed to provide a solution. The problem was eventually solved by the $SU(3)$ group, which is equivalent to rotations in a three-dimensional complex space!

How I got to the eightfold way – the story of Yuval Ne'eman. In the years 1955–57 I was serving as the Deputy Head of the Intelligence Branch at the Israel Defence Forces HQ. In 1957, when the land had rested from war I asked the then Chief of Staff, General Moshe Dayan, for leave, so that I could pursue my studies in physics and thus realize a long-standing dream. (I had earned a BSc and a Diploma of Engineering at the Haifa Technion, back in 1945). Dayan suggested that I could serve as a defence attaché in London and combine the service with studies. I accepted with alacrity and arrived in London at the beginning of 1958. I wanted to study the theory of gravitation (general relativity) which had attracted me as a student, and found out that the place for that was King's College, with the astrophysicist Bondi. However, this institution was located in a region of

supermultiplets was known in its entirety. The discovery of new particles that fitted into the gaps in the scheme was one of the most impressive successes of the model.

In 1961 only seven semi-stable spin 0 mesons were known. The η^0 had not yet been discovered. Both the eightfold way and Sakata's model predicted the existence of this particle, and the eightfold way also predicted that its mass should be about 570 MeV. Within a few months the η^0 was discovered. The mass of the particle (550 MeV) was close to the prediction of the theory. That same year (1961) the accelerators at Berkeley, Brookhaven and CERN all found mesonic resonances of spin 1, which fitted perfectly into the octet represented in Fig. 9.3, and in 1964 mesons with spin 2 which fitted into another octet were discovered.

A decisive test between the eightfold way and the Sakata model was the spin of the xi particles (Ξ^0, Ξ^-) which was unknown at that time. The eightfold way predicted a spin of $\frac{1}{2}$, like the rest of the baryons in the octet, while the Sakata model predicted $\frac{3}{2}$. In 1963 the spin of the xi was measured and found to be $\frac{1}{2}$. However, the greatest success of the eightfold way was its prediction of the decuplet of baryons of spin $\frac{3}{2}$, culminating dramatically in the discovery of the omega minus (Ω^-).

In 1961 four baryons of spin $\frac{3}{2}$ were known. These were the four resonances Δ^-, Δ^0, Δ^+, Δ^{++}, which had been discovered by Fermi in 1952. It was clear that they could not be fitted into an octet, and the eightfold way predicted that they were part of a decuplet or of a family of 27 particles. A decuplet would form a triangle in the S–I_3 plane, while the 27 particles would be arranged in a large hexagon. (According to the formalism of $SU(3)$, supermultiplets of 1, 8, 10 and 27 particles were allowed.) In the same year (1961) the three resonances $\Sigma(1385)$ were discovered, with strangeness -1 and probable spin $\frac{3}{2}$, which could fit well either into the decuplet or the 27-member family.

At a conference of particle physicists held at CERN, Geneva, in 1962, two new resonances were reported, with strangeness -2, and electric charges of -1 and 0 (today known as the $\Xi(1530)$). They fitted well into the third course of both schemes (and could thus be predicted to have spin $\frac{3}{2}$). On the other hand, Gerson and Shulamit Goldhaber reported a 'failure': in collisions of K^+ or K^0 with protons and neutrons, one did not find resonances. Such resonances would indeed be expected if the family had 27 members. The creators of the eightfold way, who attended the conference, felt that this 'failure' clearly pointed out that the solution lay in the decuplet. They saw the pyramid of Fig. 9.4 being completed before their very high traffic, and I soon realized that it would be quite impractical to work at the embassy in the west of London, and study at a college in the east of the city. So I settled for Imperial College, which was a 5-minute walk from the embassy.

My choice of a supervisor for my thesis was somewhat incidental. I consulted the college's prospectus and approached one of the professors whose names appeared there under 'Theoretical Physics'. I told him about my interest in Einstein's unified field theory, to which he replied that he did not know if anyone was still working on that, but Abdus Salam and his group in the Mathematics Department were working on field theory. I came to Salam, presenting the only letter of recommendation that I had, which was from Moshe Dayan. Salam laughed and commented 'What can a general know about scientific abilities?' Nevertheless, he agreed to take me on probation, on the strength of my degree from the Technion – Israel Institute of Technology – and the recommendation letter. I had missed the first term, and had to absent myself from some lectures in the second term too, when my duties as a defence attaché interfered. Eventually in May 1960 I resumed my studies with the status of an officer on leave with a one-year scholarship from the Government.

I was captivated by group theory, which I had first met in Salam's course. I did not know at

eyes. Only the apex was missing, and with the aid of the model they had conceived, it was possible to describe exactly what the properties of the missing particle should be! Before the conclusion of the conference Gell-Mann went up to the blackboard and spelled out the anticipated characteristics of the missing particle, which he called 'omega minus' (because of its negative charge and because omega is the last letter of the Greek alphabet). He also advised the experimentalists to look for that particle in their accelerators. Yuval Ne'eman had spoken in a similar vein to the Goldhabers the previous evening, and had presented them in a written form with an explanation of the theory and the prediction. (The Goldhabers were graduates of the Hebrew University of Jerusalem, and Gerson – today at Berkeley – is a brother of Maurice Goldhaber, who, with Chadwick, discovered in 1934 that the deuterium nucleus contained a proton and a neutron. See section 1.3.) Gell-Mann and Ne'eman had never met before that conference, but from then on they became close friends.

These are the properties predicted by the eightfold way for the particle at the apex of the pyramid. It would be a single particle, whose position in the S–I_3 plane implied a strangeness of -3, and charge -1. Being alone at the edge of the triangle meant an isospin of zero. Like the rest of the decuplet its spin would be $\frac{3}{2}$ and its parity positive. The mass formula in this case is quite simple: the mass of the particles in the decuplet has to increase in roughly equal intervals of about 150 MeV with each unit of strangeness. Thus the mass of the missing particle was predicted to be about 1680 MeV. Despite its being in a group of resonances, and having the largest mass in this group, the omega minus was expected to be a semi-stable particle, with an average lifetime of about 10^{-10} seconds, since its high strangeness would prevent it from decaying by the strong interaction (there are no particles whose sum of strangeness is -3, and sum of mass less than the expected mass of the omega minus).

Never had the particle hunters had such a perfect identikit of a wanted particle. Fortunately they also had the tools to trap it. The new synchrotron at the Brookhaven National laboratory in New York State had just come into operation and was capable of accelerating protons to an energy adequate to produce the heavy particle, as was the accelerator at CERN.

In November 1963 a group of 33 physicists embarked on the hunt at Brookhaven. Protons accelerated to 33 GeV hit a target and produced 5-GeV kaons. These were shot into a 3-metre-diameter bubble chamber – the largest in the world at that time – filled with liquid hydrogen. Stereoscopic cameras photographed the paths of the particles produced in the collisions of the kaons with protons in the chamber. Out of about 100 000 photographs taken during the experiment, one indeed showed the clear stamp of the particle being sought (see Figs. 9.5 and 9.6). In February 1964 the success of the project was officially announced. In the dozen years that followed, a few tens of Ω^- events were recorded, as well as their anti-particles $\bar{\Omega}^+$. All the quantum numbers were found to fit the theoretical predictions. The mass was 1672 MeV, and the spin $\frac{3}{2}$. (In 1964 Yuval Ne'eman was working at Caltech, and Richard Feynman used to tease him by opening the door of his office and saying, 'Have you heard? It turns out that the spin of the Ω^- is only $\frac{1}{2}$!') One can see the progress in the field of accelerators when learning that in 1984 there are already beams of omega minus particles, with about 100 000 photographs of the particle in one experiment!

The eightfold way celebrated a victory. With the aid of its diagrams many numerical results were predicted which were later verified experimentally. However, the accomplishment most remembered and acclaimed was the discovery of the omega minus. It was compared to the three missing elements in Mendeleev's periodic table, whose properties Mendeleev had been able to predict. Actually, the importance attached to a successful prediction is associated with human psychology rather than with the scientific methodology. It would not have detracted at all from the effectiveness of the eightfold way if the Ω^- had been discovered *before* the theory was proposed. But human nature stands in great awe when a prophecy comes true, and regards the realization of a theoretical prediction as an irrefutable proof of the validity of the theory. (Indeed it is believed that an unidentified track found in 1955 by Yehuda Eisenberg of the Weizmann Institute of Science at Rehovot, in a photographic plate exposed to cosmic rays, was the path of an omega minus.)

9.4 The quark model

Despite its success, the eightfold way could not be considered a complete and dynamical theory. While it was able to predict missing particles in a nearly complete supermultiplet, it could not predict the existence of whole new supermultiplets. It found relationships between the masses of the hadrons, but did not explain why the particles had those particular masses. Some of its successes seemed to imply that it did not contain the whole story; that it was a pattern reflecting some more fundamental structure of nature which still awaited discovery.

that time that in this field Israeli scientists were in the forefront, with Yoel Racah having played a key role in applying it in atomic spectroscopy, and his pupils Igal Talmi and Amos de-Shalit applying it in nuclear physics.

I learnt group theory from a textbook by A. B. Dynkin, translated from Russian (in 1978 we met in Tel Aviv, after his emigration from the USSR). It was Salam who suggested that I should study Dynkin's work. He understood that I was interested in classifying the particles and finding their symmetries, when on the basis of my rather scanty knowledge of group theory I began to suggest possible models in this area. When I first showed him my suggestions, Salam told me that they had already been attempted five or six years previously. My next proposals, he told me, had been suggested by others two or three years ago. I felt that I was getting nearer to the present . . . However, Salam was sceptical. He proposed another subject, and when I insisted on trying to classify the hadrons he said: 'I wanted to assign you an easier problem, on the assumption that it would be better for you to produce a complete work during your one year of leave from the army. However, since your mind is made up, do it your way, but you should know that you are embarking on a highly speculative search. If indeed you have already decided, at least do it properly. Learn group theory thoroughly.' So he recommended the

Dynkin book. I learnt Dynkin's work and was now able to define what I was looking for: a rank two Lie group which would permit the classification of the hadrons according to isospin and strangeness, in a manner that would fit the properties of the observed particles. I found that only four groups might do the job, and began to examine each of these separately. I remember that one of them (called $G(2)$) yielded diagrams in the shape of the Star of David, and I hoped that it might be the correct one – but it was not. On the other hand, $SU(3)$ gave a perfect fit! I finished my work in December 1960, and discussed it with Salam. I then submitted a paper for publication early in February 1961, and it appeared shortly after. Murray Gell-Mann of Caltech was working on the same problem at the same time and had reached the same conclusion.

In the 1960s, attempts were made to combine multiplets of particles into even larger groups than the supermultiplets of the eightfold way. These attempts yielded a number of successful predictions, but eventually reached a dead end.

The big breakthrough came when there was a return to the discarded notion that the hadrons were composed of a small number of fundamental building blocks. The revival of this idea was rooted in the eightfold way; actually it was hidden in its diagrams, much as the key to the structure of the atom lay in Mendeleev's periodic table.

The authors of the eightfold way were among the first to realize the implications of their model. In 1962, Ne'eman (then Scientific Director of the Laboratories of the Israel Atomic Energy Commission) and Haim Goldberg-Ophir, who had just completed his PhD under Yoel Racah, proposed a model in which each particle in the semi-stable baryons octet (Fig. 9.2) was made up of three fundamental components, each possessing a baryon number of $\frac{1}{3}$. A paper was sent to the journal *Il Nuovo Cimento*, where for a time it seems to have been mislaid, but eventually it was published in January 1963. The theory did not attract much attention, both because the eightfold way had not yet won general recognition (the Ω^- had not yet been discovered) and also because it did not go far enough. The authors had developed the mathematics resulting from the eightfold way, but they had not yet decided whether to regard the fundamental components as proper particles or as abstract fields that did not materialize as particles.

In 1964 a more explicit and detailed presentation was published by M. Gell-Mann and George Zweig, also of Caltech, who worked independently of each other. According to this model, all of the hadrons then known were composed of three fundamental particles (and their three anti-particles) which Gell-Mann named quarks. (This fanciful name, taken from James Joyce's *Finnegans wake*, has no meaning in English, but in German it means 'liquid cheese'.) This time the response was immediate. The omega minus had just been discovered, bringing great prestige to the eightfold way, and physicists were enthusiastic about exploring ideas connected with it. Moreover, Gell-Mann's paper, published in *Physics Letters*, was lucid and clear. It set out the special properties of the quarks, cautiously suggesting that they might exist as free particles and challenged the experimentalists to look for them.

Gell-Mann designated the quarks u, d and s; u and d standing for 'up' and 'down', the directions of the isospin, and s standing for 'strange' (since it carries strangeness). The anti-quarks were desig-

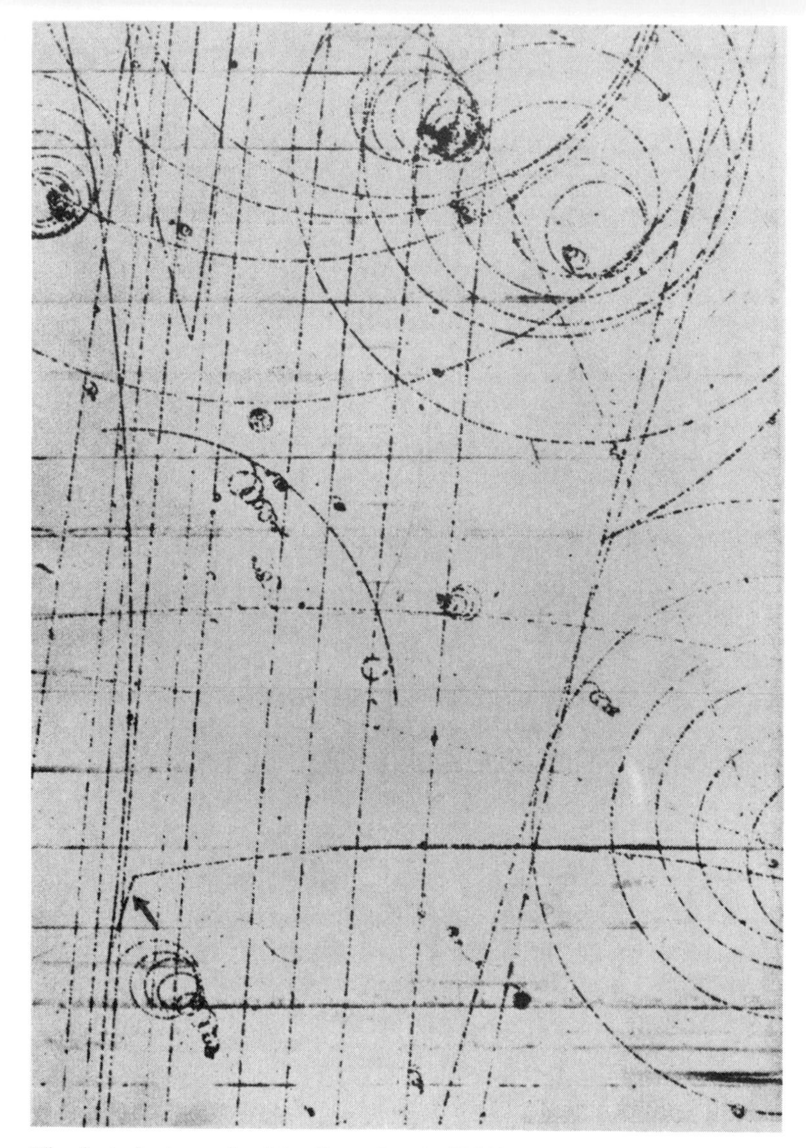

Figure 9.5. The first photograph of the Ω^-, taken in 1964 in a bubble chamber at the Brookhaven National Laboratory. The almost straight lines crossing the photograph obliquely upwards are the tracks of high-energy K^- particles. The other lines are paths of particles formed in collisions between the kaons and protons in the bubble chamber. The path of the Ω^- is marked with a small arrow at the lower left of the photograph. It was identified with certainty on the strength of its decay, in a three-stage process, into non-strange particles (see Fig. 9.6). (Courtesy Brookhaven National Laboratory, Upton, New York.)

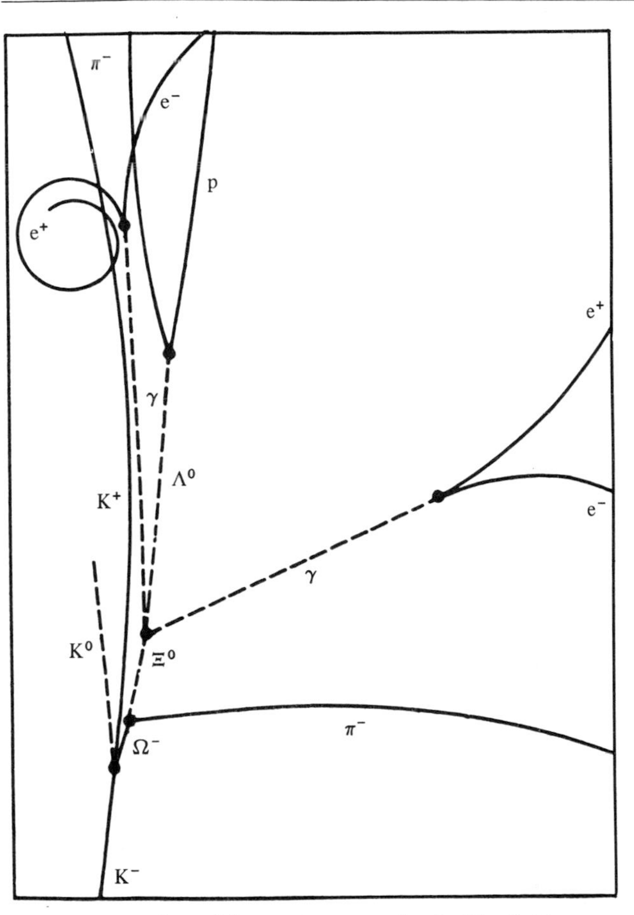

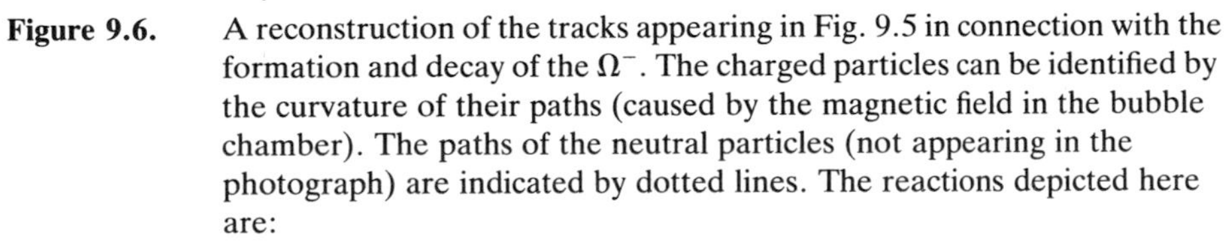

Figure 9.6. A reconstruction of the tracks appearing in Fig. 9.5 in connection with the formation and decay of the Ω^-. The charged particles can be identified by the curvature of their paths (caused by the magnetic field in the bubble chamber). The paths of the neutral particles (not appearing in the photograph) are indicated by dotted lines. The reactions depicted here are:

$$K^- + p \rightarrow \Omega^- + K^+ + K^0$$
$$\Omega^- \rightarrow \Xi^0 + \pi^-$$
$$\Xi^0 \rightarrow \Lambda^0 + \pi^0$$
$$\pi^0 \rightarrow \gamma + \gamma$$

Table 9.2. *Properties of quarks*

Quark	Electric charge, Q	Baryon number, A	Isospin	Strangeness
u	$\frac{2}{3}$	$\frac{1}{3}$	$\frac{1}{2}$	0
d	$-\frac{1}{3}$	$\frac{1}{3}$	$\frac{1}{2}$	0
s	$-\frac{1}{3}$	$\frac{1}{3}$	0	-1
$\bar{u}$	$-\frac{2}{3}$	$-\frac{1}{3}$	$\frac{1}{2}$	0
$\bar{d}$	$\frac{1}{3}$	$-\frac{1}{3}$	$\frac{1}{2}$	0
$\bar{s}$	$\frac{1}{3}$	$-\frac{1}{3}$	0	1

nated $\bar{u}$, $\bar{d}$ and $\bar{s}$. The properties of the quarks and anti-quarks are shown in Table 9.2.

One of the outstanding features of the quarks is their electric charge. Contrary to all the previously discovered particles, they have non-integral charges (in units of the proton charge): it is $\frac{2}{3}$ for u and $-\frac{1}{3}$ for d and s. The baryon number of each quark is $\frac{1}{3}$; as for strangeness, it is 0 for u and d, and -1 for s. The electric charge, baryon number and strangeness of the anti-quarks ($\bar{u}$, $\bar{d}$ and $\bar{s}$) are, of course, opposite in sign. Both the quarks and anti-quarks were assigned spin $\frac{1}{2}$. The u and d quarks form an isospin doublet ($I = \frac{1}{2}$) while the s quark is an isospin singlet ($I = 0$).

The quark model claimed that all the known hadrons were composed of quarks and anti-quarks according to the following simple rules:

(1) Each meson is a quark–anti-quark pair.
(2) Each baryon consists of three quarks, and each anti-baryon of three anti-quarks.

This simple model accounted to perfection for the properties of all the hadrons known in the 1960s. Not only could each meson and baryon be interpreted as made up of two or three quarks according to the above rules (see Fig. 9.7), but all the allowed combinations of quarks and anti-quarks correspond to known hadrons. For example, under the model, π^+ is composed of u and $\bar{d}$, arranged with their spins in opposite directions so that they cancel each other out, and with no orbital angular momentum relative to each other. Thus the spin of the system is zero. The baryon number is zero too ($\frac{1}{3} - \frac{1}{3} = 0$) and the charge is +1 ($\frac{2}{3} + \frac{1}{3}$). As for π^-, it is interpreted as $\bar{u}d$, combined in the same way. K^+ is $u\bar{s}$ (the $\bar{s}$ contributes the strangeness!), and K^0 is $d\bar{s}$. The baryons in Table 5.2 can be regarded as composed of three quarks, for example:

$$p = uud,\ \bar{p} = \bar{u}\bar{u}\bar{d},\ n = udd,\ \Lambda^0 = uds$$
$$\Sigma^+ = uus,\ \bar{\Sigma}^- = \bar{u}\bar{u}\bar{s},\ \Omega^- = sss$$

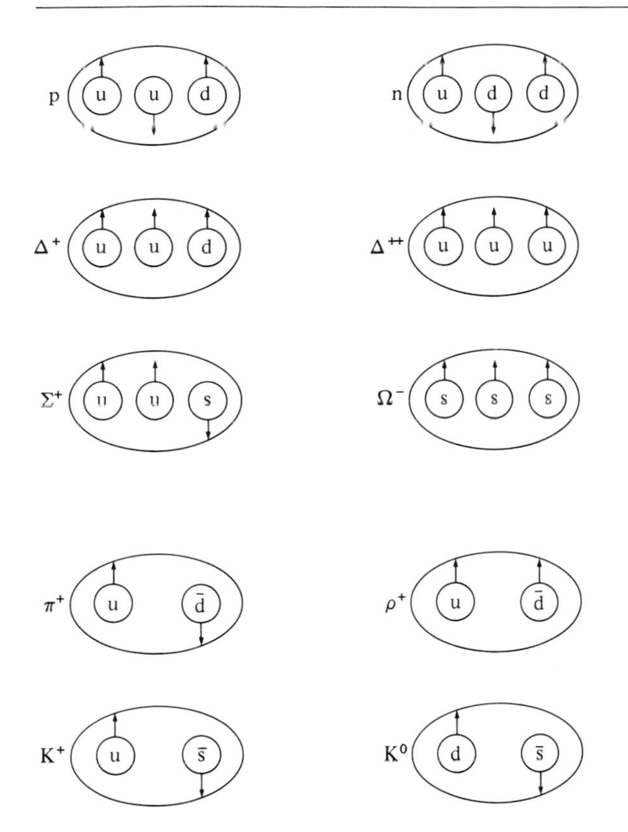

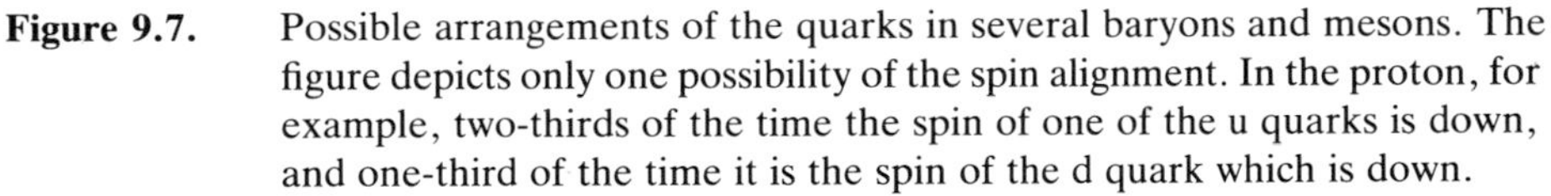

Figure 9.7. Possible arrangements of the quarks in several baryons and mesons. The figure depicts only one possibility of the spin alignment. In the proton, for example, two-thirds of the time the spin of one of the u quarks is down, and one-third of the time it is the spin of the d quark which is down.

It is easily verified that in each case the electric charge, baryon number and strangeness of the particle equals the sum of the corresponding quantum numbers of the composing quarks. It should be noted that the quark model applies to hadrons only – that is, to particles that are influenced by the strong force. The leptons – the electron, muon and their neutrinos – are *not* made up of quarks.

The quark model explained, as if by magic, many facts that had hitherto been mysteries. Why, for instance, had only nine mesons of spin zero been found at energies below 1500 MeV (namely the eight mesons in the octet of Fig. 9.1 and the η')? The answer was simple: exactly nine different pairs consisting of a quark and anti-quark could be formed from three quarks and three anti-quarks: $u\bar{u}$, $u\bar{d}$, $u\bar{s}$, $d\bar{u}$, $d\bar{d}$, $d\bar{s}$, $s\bar{u}$, $s\bar{d}$, $s\bar{s}$. Inserting these combinations into a table according to strangeness (S) and electric charge (Q), we find that the nine mesons occupy only seven places (see Table 9.3).

Table 9.3. *The nine combinations of quark–anti-quark pairs, arranged according to strangeness (S) and electric charge (Q)*

Q \ S	1	0	−1
1	u$\bar{s}$	u$\bar{d}$	
0	d$\bar{s}$	u$\bar{u}$, d$\bar{d}$, s$\bar{s}$	s$\bar{d}$
−1		d$\bar{u}$	s$\bar{u}$

The quark model thus explained why there was no meson of strangeness −1 and charge +1, or strangeness +1 and charge −1, a fact which had no explanation before.

In the same way it can be shown that there are just 10 different combinations of three quarks, which correspond to 10 combinations of S and Q, as shown in Table 9.4.

Again, only after the quark model had been published could one understand why a baryon of strangeness −3 and charge −1 had been discovered (the Ω^-), but not a baryon of strangeness −3 and charge 0 or +1, or a baryon of strangeness −2 and charge +1 (which according to the $SU(3)$ formalism, could have existed in a supermultiplet of 27 particles). It can also be shown that the three-quark model permits supermultiplets of 1, 8 and 10 particles (but not 27) and that the symmetry of these fits the symmetry of $SU(3)$. In other words, the quark model automatically leads to the octet and decuplet patterns of the eightfold way. On the basis of this model it is easy to understand the success of the eightfold way, just as the elucidation of the structure of the atom explained why the periodic table had been so successful.

In 1969 Gell-Mann was awarded the Nobel prize for his contribution to the classification of the fundamental particles. The citation, which emphasized the discovery of strangeness and the 'eightfold way' mentioned also the concurrent and independent work of Ne'eman. Ne'eman was awarded the 1969 Einstein prize for his contribution to the classification of particles, and particularly his proposal, inspired by the eightfold way, of a fundamental field of baryon number $A = \frac{1}{3}$, a precursor of the quark. (The Einstein prize was established in 1949, in connection with the seventieth anniversary of Einstein, and Ne'eman was the first non-American scientist to receive this prize.)

The connection between resonances and semi-stable particles is also explained – in principle at least – by the quark model. For example, according to the model the particle Σ^+, whose mass is 1190 GeV and spin $\frac{1}{2}$, is composed of the quarks uus. Also the resonance $\Sigma^+(1385)$, with spin $\frac{3}{2}$, consists of uus. The difference is that in the first case two of the spins cancel each other, while in the second case all three spins are parallel to each other and the binding energy is lower (therefore the mass is higher). Another example is the mesonic resonance ρ^+, which consists of u$\bar{d}$, as does π^+, except that in ρ^+ the spins are parallel and add up to 1, while in π^+ they are antiparallel and cancel each other.

On the subject of spins, the quark model also explains why mesons have integral spins and baryons half-integral spins. The spins of the quark and anti-quark in each meson can be parallel

Table 9.4. *The ten combinations of three quarks, arranged according to strangeness (S) and electric charge (Q)*

Q \ S	0	−1	−2	−3
+2	uuu			
+1	uud	uus		
0	udd	uds	uss	
−1	ddd	dds	dss	sss

(adding up to a total spin of 1) or antiparallel (total spin 0). The orbital angular momentum of the quarks can, as usual, be 0, 1, 2 . . . Thus the total angular momentum – which is the spin of the meson – is always an integer.

Baryons, on the other hand, consist of three quarks; their orbital angular momentum is a whole number, while the three spins add up to $\frac{1}{2}$ or $\frac{3}{2}$. The total angular momentum (which is the baryon spin) is therefore half integral.

The baryon numbers of the baryons, anti-baryons and mesons are explained in a similar way. Since we attributed to each quark a baryon number of $\frac{1}{3}$, and to each anti-quark the baryon number $-\frac{1}{3}$, the mesons, consisting of a quark and anti-quark, must have a baryon number of zero ($\frac{1}{3} - \frac{1}{3} = 0$), while the baryons (consisting of three quarks) and anti-baryons (three anti-quarks) have the baryon numbers +1 and −1 respectively.

The quark model, though still phenomenological, brought order into the chaotic world of particles to an extent that surpassed the wildest dreams of the physicists. It was also a veritable boon to students: instead of overburdening their memories with names and properties of dozens of hadrons, it was now possible to begin with the quarks, and then learn about the particles into which they built.

Note that only two quarks, u and d, are sufficient to build all the stable hadronic matter in our world. Strange particles contain at least one s (or $\bar{s}$) quark. For instance, $K^+ = u\bar{s}$, and therefore its strangeness is +1, while $K^- = \bar{u}s$ and its strangeness is −1. The composition of Ω^- is sss, and it therefore has a strangeness of −3. Actually the quark model gave a new meaning to strangeness, which hitherto had been a rather abstract quantum number. The strangeness of a particle can now be simply defined as the difference between its number of anti-quarks $\bar{s}$ and quarks s. The conservation of strangeness in strong interactions means that in those interactions an s quark cannot be transformed into a u or d quark. The conservation of baryon number is equivalent to the conservation of the number of quarks minus the number of anti-quarks.

9.5 The confined quarks

Shortly after the suggestion of the quark model, an enthusiastic search for them began. It was hoped that a collision between two hadrons, if it was violent enough, would break them up and set individual quarks free.

The search for quarks was based mainly on their fractional electric charge. Since the number of ions produced by a charged particle passing through matter is proportional to the square of the charge, the ionization power of a quark with charge $\frac{1}{3}e$ or $\frac{2}{3}e$ would be only $\frac{1}{9}$ or $\frac{4}{9}$, respectively, of that of a 'normal' particle carrying charge e and moving at the same velocity. Thus the track of a quark in a cloud or bubble chamber, or in any other detector based on ionization, would be particularly thin. Other strategies for detecting the quark were also adopted, but all the efforts were in vain. No sign of free quarks was found, either in the large accelerators or in detectors exposed to cosmic rays.

In his original presentation of the quark model, Gell-Mann had pointed out the interesting fact that if a quark were freed from a hadron, it would not be able to decay into any known particle due to the conservation of charge. The creation of a free quark would perhaps require a great amount of energy, and may occur very rarely, Gell-Mann said, but once produced, the quark would be a stable particle. It was therefore thought that there might exist small concentrations of free quarks in matter on the earth or in space, these quarks having been formed by collisions of high-energy cosmic particles with atoms in the atmosphere, and having remained undiscovered because of their extreme rarity.

Numerous samples were therefore taken from meteorite fragments, the moon's surface, dust from the upper atmosphere, and ancient rocks on the ocean bed. These were examined with sensitive instruments capable of detecting a charge as low as $\frac{1}{20}$ of the electron charge. All these experiments have yielded negative results – no reliable evidence of a quark has been found. Actually, every several years some physicist would announce the observation of a quark, but this could never be validated. In 1977, for instance, a group from Stanford University reported having found signs of an electric charge of $\frac{1}{3}$ on a number of small metal spheres. However, other researchers could not reproduce these results. Such a single irreproducible result is usually attributed to experimental error, and cannot be taken as evidence for the existence of a new particle.

Physicists have therefore concluded that the extraction of a free quark from a hadron is an extremely rare event, and may be even impossible.

The failure to find a free quark could be accounted for in two different ways. One assumes that it is possible in principle to free quarks from a hadron, but that the energies available today are not high enough to overcome the strong binding energy of the quarks. In that case, when sufficiently high energy accelerators become available in the future, it might be possible to knock a quark out of, say, a proton, by bombarding it or by smashing two protons together. Because of the link between energy and mass, this view would imply that each quark possesses a very large mass, most of which converts to binding energy within the hadron ('lucky that some mass is left over for the mass of the hadron' physicists once remarked jokingly). This explanation of our inability to discover isolated quarks, widely accepted at one time, contradicted results of experiments first performed in 1969–70 in which electrons were scattered by protons or neutrons. These experiments indicated that the quarks within the nucleons behave almost like free particles.

Another explanation, generally accepted today, is that the nature of the forces between the quarks makes their release impossible in principle. It is suggested that the inter-quark force differs in principle from all the forces hitherto known, and instead of weakening with increasing distance between the quarks, it is weak or nil when the quarks are very close, but becomes rapidly stronger when the distance between the quarks increases beyond the diameter of the proton. Therefore, in high-energy collisions between protons or protons and electrons, where the colliding particles actually penetrate each other, the quarks behave as free particles. However, in order to remove a quark over a macroscopic distance away from the other quarks in a hadron, an extremely high energy would be needed. Yet even if such high energy could be supplied to a proton, for instance, it would not produce a free quark. The reason is that when the applied energy becomes higher than two quark masses, it would materialize into a new quark–anti-quark pair. The new quark would take the place of the one extracted from the proton, while the new anti-quark would cling to the extracted quark, to form a meson (say, a pion). So instead of a free quark, a free meson would be produced.

According to this explanation, free quarks cannot be produced, no matter how much energy we supply. The quarks are forever confined within the hadrons, and cannot escape.

Several vivid analogies of this situation have been given. Accord-

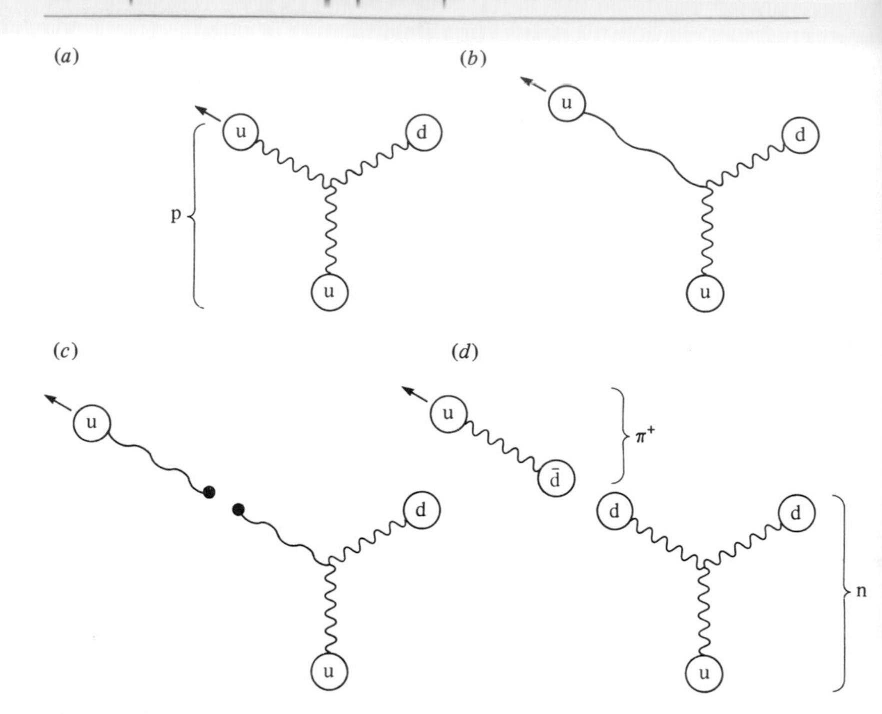

Figure 9.8. According to the string model, when we try to draw a quark out of a hadron by breaking the connecting 'string', a new quark and anti-quark pair appears at the broken ends. In the case of a proton, the observed reaction might be: $p \rightarrow n + \pi^+$. (*a*)–(*d*) describe very schematically several stages of this process.

ing to one, the quarks in the hadrons are joined to each other by a sort of elastic string, which is actually a force field in which the lines of force do not spread out in all directions as in an electric or gravitational field; they stretch straight from quark to quark. When one tries to separate the quarks by breaking the string, a new quark and anti-quark are formed at the broken ends (see Fig. 9.8). Another analogy regards the quarks in the hadron as being in a sort of elastic bag which cannot be broken and within which the quarks move like almost free particles.

According to the described model of the inter-quark force, the so-called 'strong force' between hadrons is only a pale shadow of the 'really strong force' between quarks, just as the van der Waals' forces between neutral molecules are the puny remnant of the electric force.

As mentioned above, when the quarks are very close to each other within the hadron, the force between them becomes very weak, and they behave as free particles. Under these circumstances the binding energy of the quarks is small, and it is not necessary to assign high masses to them. According to one approach, the effective masses of the u and d quarks would be about 0.34 GeV (about a third of a nucleon's mass) and that of the s quark about 0.54 GeV. These values are based on the assumption that the masses of certain hadrons (such as the nucleons and hyperons) are given by the sum of the masses of their constituent quarks. The assumption that the mass of s is greater than that of u and d is based on the fact that the greater the strangeness of a hadron (and the more s quarks it has) the greater its mass. However, looking in depth at modern theory we find that when a u or d quark floats about in the middle of the hadron, its effective mass is very small, of the order of the electron mass. The nucleon's mass then mostly reflects the tremendous potential energy carried by the carriers of the string-like forces between the quarks.

The failure to discover free quarks, on the one hand, and the impressive successes of the quark model on the other, sparked a debate on the significance of the model. In the 1960s and early 1970s, some important physicists opposed this model vehemently. For example, Werner Heisenberg, Nobel prize laureate and the grand old man of quantum mechanics, declared in 1976 at a conference of the Physical Society of Germany that the question of what the elementary particles were composed of was a philosophical question of no physical significance, and it would be better not to deal with it.

Some of the arguments of the opponents of the quark theory are worth discussing here. When the atomic nucleus is bombarded with particles with energies of several MeV, and neutrons and protons are ejected out of it, we can say that those neutrons and protons were formerly part of the bombarded nucleus. True, there is a certain inaccuracy, because the nucleons inside the nucleus differ slightly from the freed nucleons – for one thing, their mass is slightly lower inside the nucleus. Nevertheless, the model which regards the nucleus as a system of Z protons and $A - Z$ neutrons is universally accepted, because of its remarkable success in explaining experimental observations. When it comes to shattering the shell of a particle like the proton, much higher energies – of the order of several GeV – would be required. In collisions at such energies, a great deal of kinetic energy would be converted to mass, and one would not be able to say whether the products of the reaction were the constituents of the original particle. In a collision between protons, for example, pions are generated. Can we conclude from that observation that protons are composed of pions? That would be a rather hasty conclusion, since in high-energy collisions between pions, protons could be formed! How could we then determine which particles are composed of which? Obviously, we must be careful when making an assertion about one sub-atomic particle being built up from other particles. Owing to the conversion of energy to mass, the phrase 'composed of' would seem to have a different meaning from its sense in the statement that 'a car is composed of an engine, body and wheels'.

The opponents of the quark model thus argued that until the detection of a free quark occurred there was no point in talking about quarks at all, and even if quarks could be liberated from a particle, it did not mean that the particle was 'composed' of those quarks.

It may be mentioned here that in the absence of an adequate dynamic theory for describing the behaviour of particles, a model was proposed in the 1950s that called for 'democracy' rather than hierarchy in the world of particles. In this model each hadron was regarded as some combination of all the others. Ne'eman used to depict this approach by the following analogy. Think of a jungle in which tigers, lizards and mosquitoes coexisted. The tigers fed on the lizards, the lizards fed on the mosquitoes, and the mosquitoes stung the tigers and sucked their blood. One could therefore say that the tigers were 'composed' of lizards, the lizards of mosquitoes and the mosquitoes of tigers. The view that each particle was 'composed' of all the others was dubbed 'the bootstrap model' because, as in the famous story of Baron Münchausen, the only way out of the morass was to pull yourself up by your own bootstraps. This approach was popular in the 1960s, as long as it seemed impossible to treat the strong interactions with the same theoretical methods that had proved so successful in quantum electrodynamics.

One of the stalwarts of the bootstrap model was Geoffrey Chew of the University of California at Berkeley. In the years 1955–73 Chew, a charismatic figure, was influential in spreading new ideas in physics with almost religious fervour. The success of the quark model depressed him greatly. His lectures at a summer school in the early 1970s were entitled: 'Bootstrap or quarks – triumph or tragedy?'

9.6 Experimental evidence for the quark model

Yet the opposition to the quark model could not but retreat in the face of its numerous impressive successes, concerning a wide range of experimental observations. It explained not only the success of $SU(3)$, but many phenomena that had no connection with $SU(3)$. We have mentioned that the quark model provided a mechanism explaining the connection between resonances and semi-stable particles. It also made it possible to calculate the magnetic moments of various hadrons, and to predict the results of interactions between particles. For instance, according to the model the ratio between the probability of interaction between nucleons and a target, compared with that of pions and the same target, should tend to $\frac{3}{2}$ at high energies, since at high energies the probability of interaction should be proportional to the number of quarks (and anti-quarks) in the particle. This ratio was indeed verified by experiment.

Convincing evidence of an internal structure in the proton was provided by experiments on the scattering of fast electrons by protons which have been conducted at SLAC (Stanford) and DESY (Hamburg) since the late 1960s. In these experiments a stationary target was bombarded with electrons with energies of 20 GeV or more. When such a high-energy electron hit a proton head on, it was able to penetrate it and probe its inner structure. It was found that most of the electrons that passed through the target underwent only small deviations from their original path. However, the number of large-angle deviations was significantly larger than what would have been expected if the proton mass and charge were uniformly 'spread' throughout its volume. These results are reminiscent of Rutherford's experiment which probed the atom and revealed its nucleus (see section 1.1). Indeed, analysis of the SLAC and DESY results showed that they were consistent with most of the proton being electrically neutral with just several charged 'nuclei', very small in comparison with the proton's diameter. Richard Feynman of Caltech initially named these objects 'partons'. They turned out to have spin $\frac{1}{2}$, and could be the captive quarks.

Experiments on the scattering of muons by protons and on interactions of neutrinos with protons conducted at CERN and Fermilab gave similar results. The interaction of a low-energy muonic neutrino (ν_μ) with a proton yields a neutron and a positive muon. But when the incident neutrino has an energy of several tens of GeV, a large number of particles may be produced in a single reaction, among them various hadrons. Bubble chamber photographs of such reactions revealed some cases where a number of hadrons were emitted in a narrow beam, or jet, which made a wide angle with the direction of the incident neutrino. Analysis of these events showed that they could be interpreted as a collision between the neutrino and a quark inside the proton. As a result of that collision the quark is ejected, but immediately new quarks and anti-quarks are created between it and the proton that has been hit. The observed result is a jet of hadrons, all moving in approximately the same direction.

The electron, muon and neutrino are regarded as point particles, and so analysis of their (weak and electromagnetic) interactions with nucleons, in terms of quarks (which can thus also be regarded as pointlike) is relatively simple. The analysis of high-energy collisions between two protons, which involve the (strong) interaction between six quarks is more complicated. Nevertheless, such experiments too hinted unambiguously at some inner structure of the proton, and ruled out the possibility of its being a uniform sphere, like a ball of cotton wool. For example, in experiments conducted since 1972 at the intersecting storage rings at CERN, collisions between two beams of protons at a total energy of up to 63 GeV were investigated. In such a collision, part of the kinetic energy of the protons is converted into a number of created particles, which are usually ejected in directions close to those of the original beams. Occasionally, however, a jet of particles emerges perpendicular to the beam directions. Such jets, like the jets in neutrino–proton reactions, indicate the presence of something small and hard inside the proton. In fact, the jet is interpreted as the result of a head-on collision between a quark in one proton and a quark in the other proton.

Similar jets of particles were observed in a completely different type of experiment: annihilations of electrons and positrons. In the past, these annihilation processes exhibited impressive evidence of the equivalence of mass and energy. In this process all the mass of the colliding particles is transformed into pure electromagnetic energy, a sort of virtual photon of high energy and zero momentum (like the total momentum of the electron–positron pair before the collision) which after a very short time turns into real particles.

In the 1970s, when e^+e^- storage rings opened the way for annihilations at energies in the GeV range, there was renewed interest in these processes. At these energies not only photons are formed, as in low-energy annihilations, but also various kinds of leptons and hadrons. Since all the quantum numbers of the original particles are obliterated in this violent merging of matter and anti-matter, the formation of new particle–anti-particle pairs ensuing after the annihilation is not limited by any conservation law if only sufficient energy is available. The whole process is mainly electromagnetic, and was analysed by quantum electrodynamics, which is regarded as one of the most successful of physical theories.

The study of high-energy annihilation events in storage rings such as SPEAR at SLAC (Stanford) and Doris at DESY (Hamburg) showed that a common consequence of an annihilation is the creation of a particle and anti-particle, such as a new electron–positron pair, or a pair of muons, $\mu^+\mu^-$. But in many cases the observed products are a number of hadrons, concentrated in two narrow jets moving in opposite directions. It was concluded that the initial products of the annihilation are a quark and anti-quark, which, as they move apart, form a train of new quark–anti-quark pairs, which eventually combine to form the hadrons which reach the detectors. Due to the conservation of momentum, all the hadrons produced by the initial quark continue to move in the direction of that quark, while those produced by the initial anti-quark follow the direction marked by that anti-quark. This would explain why the detected hadrons are concentrated in two jets moving in opposite directions, instead of spreading out in all directions. The angles between the jets and the original beams of electrons and positrons indicate that the spin of the quarks is $\frac{1}{2}$.

Finally we shall mention another series of relevant experiments which was conducted in 1979 at Fermilab, near Chicago. In these experiments nucleons were bombarded with pions of energies in the 10 GeV range. Among the products of the πp interactions, pairs of muons, $\mu^+\mu^-$, were found. It turned out that when the bombarding pions were negative, four times as many muon pairs were formed than when the pions were positive. It was also found that in proton–proton collisions at the same energies, the production of muon pairs was a much rarer event.

These results are inexplicable if the quark model is ignored. The quark model, however, provides the following simple and elegant explanation. The muons are produced by annihilation of a quark in the proton and an anti-quark in the pion. Since protons contain no anti-quarks, such annihilations cannot occur in proton–proton collisions, and therefore formation of muon pairs in this process is very rare. As for the ratio of 4 : 1 between the muon formation rate from π^-p and π^+p collisions, there is a basic rule according to which the probability of high-energy annihilations is proportional to the sum of squares of the electric charges (of the colliding particle and anti-particle). In the collision of a positive pion ($\pi^+ = u\bar{d}$) with a proton (uud), the annihilation involves a d in the proton and $\bar{d}$ in the pion, whose charges are $-\frac{1}{3}$ and $\frac{1}{3}$ respectively. The probability of annihilation is therefore proportional to $\frac{2}{9}$:

$$(-\tfrac{1}{3})^2 + (\tfrac{1}{3})^2 = \tfrac{2}{9}$$

In the case of the negative pion ($\pi^- = \bar{u}d$) the annihilation involves u and $\bar{u}$, whose charges are $\frac{2}{3}$ and $-\frac{2}{3}$, and the probability of the process is proportional to $\frac{8}{9}$:

$$(\tfrac{2}{3})^2 + (-\tfrac{2}{3})^2 = \tfrac{8}{9}$$

The ratio of probabilities is thus $\frac{8}{9}:\frac{2}{9} = 4$, exactly as found experimentally.

To sum up, the quark model successfully explains hundreds of experimental results. In the experiments the quarks usually behave as if they are floating almost freely in the hadron, and the bonds between them are even weaker than between nucleons in the nucleus. Nevertheless, nobody has yet succeeded in extracting a quark from a hadron. This has been explained by assuming that the forces between the quarks are feeble when they are close together, but become very strong when the distance between the quarks increases. The result is a phenomenon called 'the confinement of quarks'; the quarks are doomed to stay forever within the hadrons, and cannot be found as free particles.

The various experimental results have greatly strengthened the standing of the quark theory. So long as the evidence consisted only of the success of that theory in explaining the properties of the various hadrons, one could still regard the quarks as fictitious mathematical tools, and reserve judgment as to their existence as real entities. But once the scattering experiments showed that inside the nucleons there were indeed small hard objects of spin $\frac{1}{2}$, it became generally accepted that quarks are real, and actually do exist in the hadrons. Despite the fact that no one has yet detected the tracks of released quarks experimentally, it is today difficult to deny their objective existence, or consider them a mere mathematical exercise.

9.7 Coloured quarks

Shortly after Gell-Mann and Zweig presented the quark model, physicists pointed out that under this model two or three quarks in a hadron could be in the same state, i.e. could have identical sets of quantum numbers. For example, according to the model, the Ω^- consists of three s quarks: $\Omega^- = \text{sss}$. Analysis of the properties of that particle, which possesses spin $\frac{3}{2}$, shows that all three quarks have parallel spins, and zero orbital angular momentum. The particles Δ^{++} and Δ^- are regarded as the ground state of the quark combinations uuu and ddd, and here too, the three quarks have parallel spins and are in identical physical states. This is surprising, since the quarks have spin $\frac{1}{2}$, which means that they are fermions, and one would expect them to obey the Pauli exclusion principle, which states that no two identical fermions co-existing in the same system can have the same set of quantum numbers.

Faced with this dilemma, physicists had several options. One option was to assume that quarks do not obey the Pauli principle. It was suggested by O. W. Greenberg of the University of Maryland that quarks obey new 'para-fermi' statistics, allowing any three quarks – but no more than three – to have the same quantum numbers. Another explanation was suggested by two physicists at the University of Chicago, Y. Nambu and M. Y. Han. According to this explanation, the quarks do obey the Pauli principle, but they possess an additional quantum number, which can have three different values for each quark. In the Ω^-, for example, the three s quarks are indeed identical in all the familiar quantum numbers (strangeness, electric charge, spin direction, etc.) but differ with respect to the new quantum number. The additional quantum number characterizing the quarks was arbitrarily named 'colour' (this term has, of course, nothing to do with the everyday meaning of the word colour). The three values of the colour are usually referred to as red (R), yellow (Y) and blue (B). Each of the quarks we have encountered can appear, according to this explanation, in three versions. For example: red u, yellow u and blue u. The three s quarks in the Ω^- are a red s, a yellow s and a blue s. Anti-quarks are assigned opposite colours: 'anti-red', 'anti-yellow' and 'anti-blue'.

A combination of the three different colours (or the three anti-colours) is colourless, or white, and so is a combination of a colour and its corresponding anti-colour. An important principle of the colour model is that all mesons and baryons are always colourless, i.e. made of combinations whose overall colour is zero or white. Thus, the three quarks in a baryon are always of three different colours, even when it is not required by Pauli's principle. Similarly, the quark and anti-quark in a meson are of two opposite colours, so that mesons too are colourless. Since all the observable particles are colourless, the colour, unlike other quantum numbers (baryon number, strangeness, etc.) cannot serve as a basis for classifying the hadrons, nor has it a direct influence on interactions between them.

9.8 Quantum chromodynamics and gluons

The notion of colour was originally introduced quite arbitrarily to explain what appeared to be a contradiction to the Pauli principle. However, in due course it has proved very useful to other aspects of the quark theory as well.

According to the current theory of the strong force, the force between the quarks in the hadron originates in the colour. In other words, the colour is the source of the inter-quark strong force. This is mathematically analogous to the 'unitary spin' charges (the group is again $SU(3)$ with eight charges although it is here on a different footing). There is also a real analogy to the electric charge being the source of the electromagnetic force. The concept of colour explains why the force acting between quarks is so different in character from the force between hadrons: the 'really strong' force acts only between coloured particles, whereas the hadrons are colourless.

The force between particles possessing colour has been named the chromodynamic force, and the theory of colour is called *quantum chromodynamics* (QCD), after the Greek word '*chroma*' meaning colour. The theory assumes that the chromodynamic force is transmitted by eight electrically neutral particles, of zero mass and spin 1, called gluons (from the word 'glue'). The gluons in some ways resemble the photon, which transmits the electromagnetic force, but unlike the photons, which themselves have no electric charge, the gluons do have colour. More precisely, each gluon carries a combination of a colour and an anti-colour. Being themselves coloured, the gluons exert chromodynamic forces on each other. This explains their ability to serve as a 'string' binding the quarks and to cause the force to increase as the separation between the quarks increases.

In each process of emission or absorption of a gluon, the colour of the quark may change, depending on the colour and anti-colour carried by the gluon. For instance, when a blue u quark emits a gluon carrying the blue and anti-red colours (designated $G_{b\bar{r}}$), it becomes a red u quark. A red d quark absorbing this gluon will become a blue d quark:

$$u_b \rightarrow u_r + G_{b\bar{r}}$$
$$d_r + G_{b\bar{r}} \rightarrow d_b$$

There are six colour-changing gluons, $G_{r\bar{b}}$, $G_{r\bar{y}}$, $G_{b\bar{r}}$, $G_{b\bar{y}}$, $G_{y\bar{r}}$, $G_{y\bar{b}}$, and two colour-preserving gluons, which can be regarded as carrying combinations of colours and their anti-colours, and will be denoted for the sake of simplicity G_0 and G_0'. The number of gluons is thus eight.

The transmission of the chromodynamic force by exchanging gluons causes the quarks in the hadron to change colours continually, as shown in Fig. 9.9. This change of colours cannot, of course, be detected, since the hadrons are always colourless.

When a charged particle is accelerated it emits photons. Similarly, it would be expected according to QCD that an accelerated quark will emit gluons. However, no experiment has succeeded in producing a free gluon. It is supposed that the gluons, like the quarks, are confined within the hadrons, and that if a gluon does get liberated it turns into known hadrons before it manages to traverse a macroscopic distance. But, despite the fact that free gluons have not been

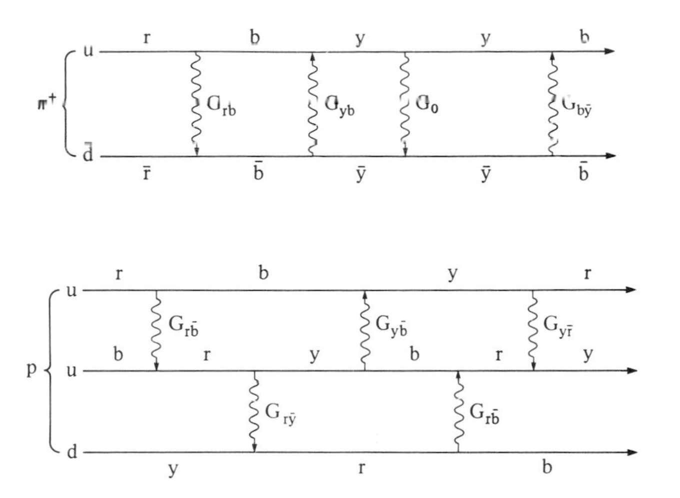

Figure 9.9. The chromodynamic force between the quarks in a meson such as π^+, or a baryon such as the proton, is transmitted by the exchange of gluons, which causes the quarks to change colours continually. The colour of the quark is designated by the letters r, b, y, $\bar{r}$, $\bar{b}$, $\bar{y}$.

detected, indirect evidence for their existence has been accumulating in recent years.

One confirmation of the existence of gluons was obtained in the scattering of electrons by protons. Analysis of the results showed that the three quarks in the proton carry only about half its momentum. The rest is apparently carried by electrically neutral, massless particles which do not affect the path of the electron. The gluons, which according to QCD are continually exchanged between the quarks, fit this description well. A similar result was obtained in the scattering of pions – the quark and anti-quark were found to carry about 40 per cent of the total momentum of the pion, the rest being apparently carried by the gluons running to and fro within it.

Other convincing evidence for the existence of gluons came from high-energy electron–positron annihilations. As we have mentioned, when hadrons are produced in such an annihilation, they tend to concentrate in two cone-shaped beams or jets. The higher the energy of the colliding particles, the narrower the cones. From a study of these events it is concluded that initially a quark–anti-quark pair is formed. As the two separate, they produce a train of additional quarks and anti-quarks, which finally coalesce to form hadrons. QCD predicted that at high enough energies – above 20 GeV – one of the initial quarks may emit some of its energy as a

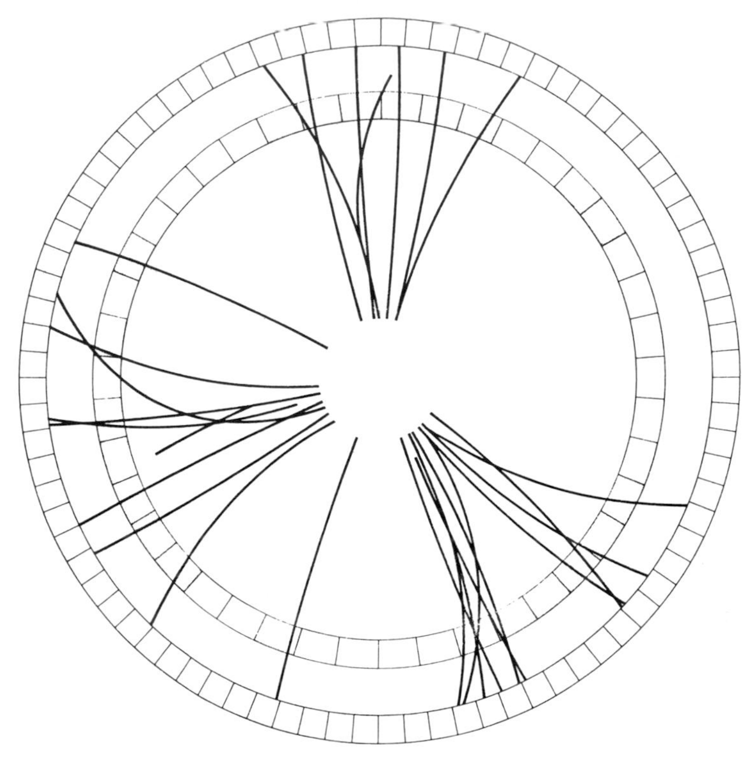

Figure 9.10. A computer reconstruction of a three-jet event recorded at the PETRA e^+e^- storage ring at DESY, Hamburg.

A resonance at 1.42 GeV which was discovered at the SPEAR storage ring (at SLAC) in 1980 and other resonances at higher energies are, somewhat speculatively, thought to be bound states of two gluons. This system is dubbed glueball or gluonium.

gluon. The gluon, like the quark, cannot proceed very far as a free particle, but will produce its own jet of hadrons, so that the detectors will record three jets instead of just two. Towards the end of 1979 a group of physicists working with the large storage ring PETRA (at DESY, Hamburg) found the first three-jets event. Later, many such events were discovered as well as four-jet events which were interpreted as gluon emission by both original quarks.

At present QCD appears to provide a good description of the 'real' strong interaction that occurs between the quarks themselves. All inter-hadron strong forces are, according to this theory, just 'the foam over that deep ocean', residual effects of the original strong force. However, there is as yet no complete theoretical proof that quark (or colour) confinement is absolute, although theorists have devoted several thousand man-years to the search of such a proof.

Chapter Ten

More quarks – or charm, truth and beauty

10.1 The theory calls for a fourth quark

At first sight, the introduction of the notion of colour tripled the number of quarks, but one could regard the 'yellow', 'red' and 'blue' u quarks as different states of the same quark, and go on talking of three quarks only – u, d and s, or quarks of three 'flavours' as physicists called them. However, it was not long before the three flavours were joined by a fourth. The existence of a fourth quark was mooted by S. L. Glashow of Harvard and J. D. Bjorken of Stanford shortly after Gell-Mann and Zweig presented their quark model, even though three quarks explained to perfection all the hadrons known in the 1960s. The first arguments for the existence of the fourth quark were purely aesthetic. Prior to the advent of the quark model it was known that the hadrons were very different from the leptons. The hadrons were numerous, of diameter 10^{-13} centimetres approximately, and they had an internal structure, as evidenced by scattering experiments. The leptons, on the other hand, were few (until the 1970s only four leptons were known), they were point particles, and apparently truly fundamental. All the leptons had spin $\frac{1}{2}$, while among the hadrons one could find particles of spin $0, \frac{1}{2}, 1, \frac{3}{2}$, etc. When the quark model was put forward, it immediately became apparent that the truly fundamental particles were the leptons and the quarks, and that these two types of particles showed surprising similarities. Both were assumed to be point particles and both had spin $\frac{1}{2}$. The similarity was marred by the fact that the leptons numbered four (ν_μ, μ; ν_e, e) and the quarks only three (d, u, s). This gave birth to the idea that a fourth quark existed which did not appear in hitherto observed particles, but might possibly appear in hadrons of larger mass.

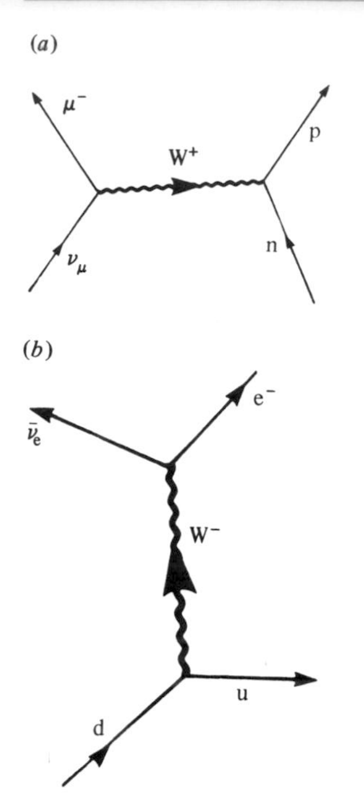

Figure 10.1. (*a*) A muonic neutrino interacts with a neutron to produce a proton and a muon. The interaction can be interpreted as follows: the neutrino emits a W^+ and turns into a μ^-. The W^+ is absorbed by the neutron and changes it into a proton, by converting a d quark into a u quark. (*b*) Beta decay can be represented as the conversion of a d quark inside the neutron to a u quark by emitting a W^-, which disintegrates into an electron and anti-neutrino. Reactions which involve the exchange of a W^+ or W^- and change the electric charge of the hadron are classed as 'charged weak currents'.

There were also more solid arguments for the existence of a fourth quark. The unified theory of the weak and electromagnetic interactions mentioned in Chapter 6 pointed to an additional symmetry between the quarks and the leptons. According to this theory, the weak force could convert ν_e to e and ν_μ to μ (and vice versa) by emission or absorption of W^+ or W^-, and could also convert a d quark to a u quark (and vice versa) in the same way, as shown in Fig. 10.1. (It could also convert an s quark to a u quark, but with a far smaller probability.) Under this scheme the s quark is left, in most cases, without a partner. To complete the symmetry a fourth quark was needed, so that there would be four quarks to match the four leptons.

During the early 1970s further evidence for the fourth quark accumulated. The most important was related to the type of re-

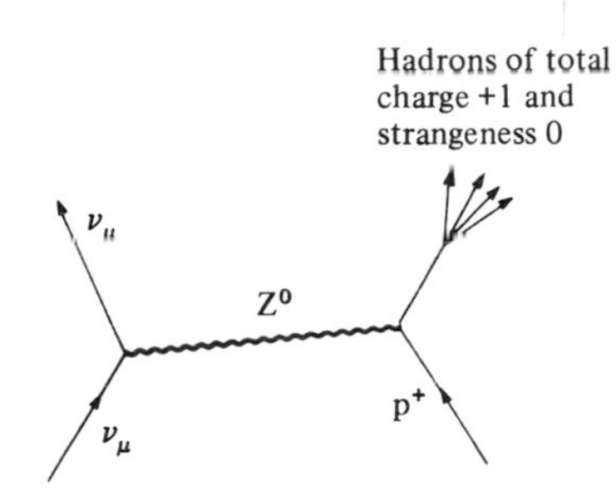

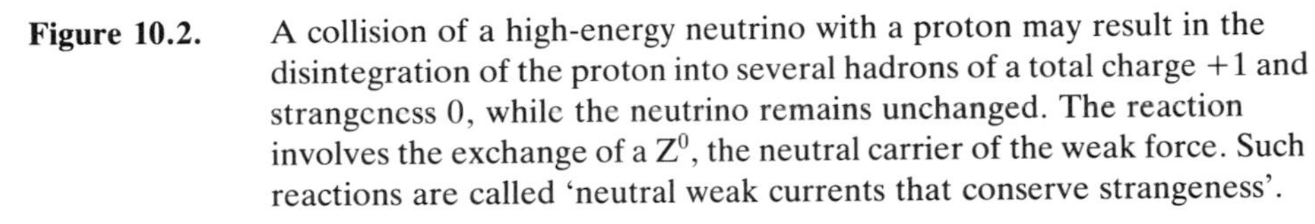

Figure 10.2. A collision of a high-energy neutrino with a proton may result in the disintegration of the proton into several hadrons of a total charge +1 and strangeness 0, while the neutrino remains unchanged. The reaction involves the exchange of a Z^0, the neutral carrier of the weak force. Such reactions are called 'neutral weak currents that conserve strangeness'.

actions known as 'strangeness-changing neutral weak currents'. These are weak interactions in which the net electric charge of the participating *hadrons* does not change, but the strangeness does. An example is the decay of a neutral kaon into a pair of muons (μ^+, μ^-). The strangeness drops from +1 to 0, but since a neutral hadron breaks up into two particles which are not hadrons, the electric charge of the hadrons taking part in the reaction is zero before and after the reaction. What bothered the physicists about the strangeness-changing neutral weak currents was the extreme rareness of these reactions. The decay of a neutral kaon into a pair of muons, for example, is very rare indeed, even though it does not contradict any conservation law and could be expected to be more frequent by a factor of 10^6. A similar type of reaction, known as 'neutral weak currents that conserve strangeness' (see Fig. 10.2) is far less rare, as was revealed by intensive research conducted in the years 1972–73.

In 1970, Glashow, J. Iliopoulos and L. Maiani of Harvard showed that the hypothesis of a fourth quark solves the problem. When K^0 (made up of the quarks $d\bar{s}$) breaks up into two muons, the $\bar{s}$ quark turns in the first stage into a $\bar{d}$ quark. If we assume the existence of a fourth quark – c, then the $\bar{s}$ has another option – to turn into a $\bar{c}$. The opening of an additional channel of disintegration creates a phenomenon similar to interference between waves and may reduce the probability of a process. It turns out that the explicit equations of the probabilities of the two processes do in fact almost cancel each other, leaving only a very small probability – if any – for the conversion of an $\bar{s}$ quark into a $\bar{d}$, in this case.

The theory provided a rather accurate description of the expected properties of the fourth quark: electric charge $+\frac{2}{3}$, strangeness 0, and a mass greater than any of the three other quarks. From data on the probabilities of various reactions, such as neutral currents, it was concluded that, like the s quark which carries the property of strangeness, the fourth quark carries a kind of flavour which no other quark has. This flavour or charge is conserved in strong and electromagnetic interactions, but not in weak interactions. The special flavour of the fourth quark was dubbed 'charm', and the quark itself 'the charmed quark', denoted by c.

10.2 A narrow resonance at 3.1 GeV

After their failure to find a free quark experimentally, physicists pinned their hopes on finding charm-carrying particles. Such a discovery could provide a firm base for the entire quark theory, since the prediction of charm sprang from that theory. But the experimentalists lagged behind the theoreticians, and for about 10 years the fourth quark and charm were hypothetical unproven concepts, as had been the case with the neutrino until 1956.

The first experimental evidence for the fourth quark appeared in the autumn of 1974, when two groups of investigators, working independently and by different methods, detected a new hadronic resonance whose properties could not be explained in terms of the three 'veteran' quarks. One group, headed by Burton Richter and Gerson Goldhaber, was working at the Stanford Linear Accelerator Center (SLAC) in California, on the study of electron–positron annihilations in the SPEAR storage ring. The second group at the Brookhaven National Laboratory, headed by the Chinese-American Samuel Ting, was investigating the particles formed in high-energy collisions of protons with a beryllium target.

The success of SPEAR was largely attributable to the advanced system of detectors set up around one of the two locations where the annihilations occurred. This system, called 'the magnetic detector', is illustrated in Fig. 10.3. It consists of a large cylindrical magnet and two types of particle detectors: spark chambers and scintillation counters (see section 1.4). Four concentric cylindrical spark chambers containing a noble gas are installed around the vacuum tube in which the electron and positron beams meet. Around these are a first series of scintillation counters, and surrounding the entire system is a giant cylindrical coil magnet of 3 metres diameter. Outside the magnet is a second series of scintillation counters and then a 20-centimetre-thick shielding of steel plates. Finally, outside the shielding is another array of spark chambers. These external detectors can only be reached by muons, since all the hadrons are stopped in the steel plates owing to the strong interaction with the

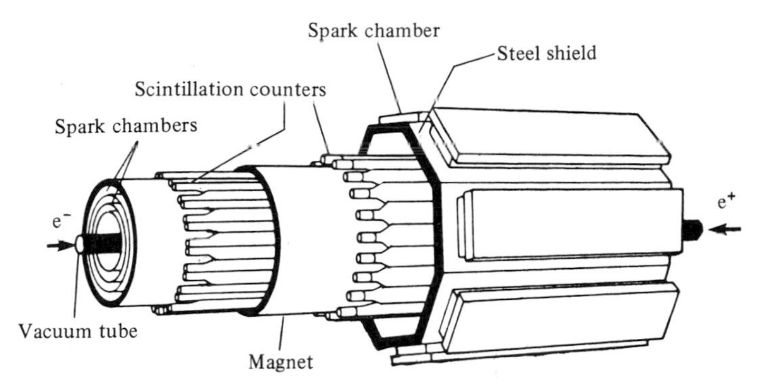

Figure 10.3. A schematic description of the 'magnetic detector' at SPEAR in California. The various layers of the detector are depicted as if they were drawn out for demonstration.

iron nuclei, while the electrons and positrons lose most of their energy in their passage through the first spark chambers and the magnet. The design of the detecting system permits immediate distinction between hadrons, muons and electrons.

It should be noted that the probability of the annihilation of an electron and positron which are moving towards each other is rather small. So although each of the particle beams counter-rotating in the SPEAR storage ring contained about 10^{11} particles, and although the beams passed through each other in the region of the magnetic detector about two million times a second, annihilation occurred only once every few seconds on average. Each annihilation activated the detectors, which recorded the trajectories of the produced charged particles moving perpendicularly to the beams – from the centre of the detector to its perimeter. The data were immediately transmitted to a computer, which sifted out the 'genuine' events resulting from annihilations from the 'spurious' events such as those resulting from collisions of electrons or positrons with air molecules remaining in the tube.

Each 'genuine' event is analysed by the computer. From the time required for the particle to pass from one detector to the next, its velocity is calculated, and from the curvature of its path in the magnetic field its momentum is obtained. From these quantities the mass is deduced. All the data are stored in the computer memory, and in the course of the experiment or later, all the particle paths can be reconstructed and displayed on a screen (see Fig. 10.8). At the end of each series of experiments, a detailed report on the number of events and their nature can be obtained.

The magnetic detector system was designed and constructed by physicists from SLAC and the Lawrence-Berkeley Laboratory. Individuals and small groups from the two laboratories were responsible for different components of the detector and the computer software connected with it. The important discoveries made with the aid of this powerful tool have made it one of the best investments ever in a physical instrument. The system eliminated the need for the extremely laborious scanning and searches which characterized work on photographs of bubble chambers. In SPEAR this hard labour was assigned to the computer. For instance, to find out how many K^- of a certain energy were produced in 100 000 annihilations, all that the physicist had to do was to strike a few keys on the computer terminal and wait a few seconds for the answer. The veteran physicists who were accustomed to the arduous visual scanning took some time to adjust to the revolution. Gerson Goldhaber of Berkeley related that at first he and his colleagues used to examine each event on the screen and record it on a piece of paper, till eventually they were convinced that the computer was better at this task and that the job should be left to it.

One of the first tasks for which the magnetic detector was used was to measure the ratio between the number of annihilations producing hadrons and those producing muons. To understand the importance of this question, we recall that in the annihilation of an electron and positron at energies in the GeV range, various particles can be formed – a new electron–positron pair, a pair of muons, or a pair of quarks which turn into two jets of hadrons (see section 9.6). The probability of formation of a particular particle in the annihilation is proportional – according to quantum electrodynamics – to the square of its electric charge. It is possible to calculate the expected ratio between the number of events in which hadrons are formed and the number of events resulting in a pair of muons. This ratio, commonly denoted by R, is equal to the sum of the squares of the charges of all the quarks that can be produced in the annihilation. Physicists were eager to measure R as a test of the quark theory. If there are three quarks whose charges are $-\frac{1}{3}$, $-\frac{1}{3}$, $+\frac{2}{3}$, the sum of the squares of the charges is $R = (-\frac{1}{3})^2 + (-\frac{1}{3})^2 + (\frac{2}{3})^2 = \frac{2}{3}$, i.e. the number of cases of hadron formation will be, on average, two-thirds of the number of pairs of muons. But if each quark occurs in three colours, the number of quarks increases threefold, and we get $R = 2$, i.e. for each muon pair there will be two events of hadron formation. At energies sufficient for the formation of the fourth quark, whose charge is assumed to be $+\frac{2}{3}$ (and assuming that the quarks have three colours), we get $R = 3\frac{1}{3}$.

The first experiments to measure R were conducted in 1973 in the

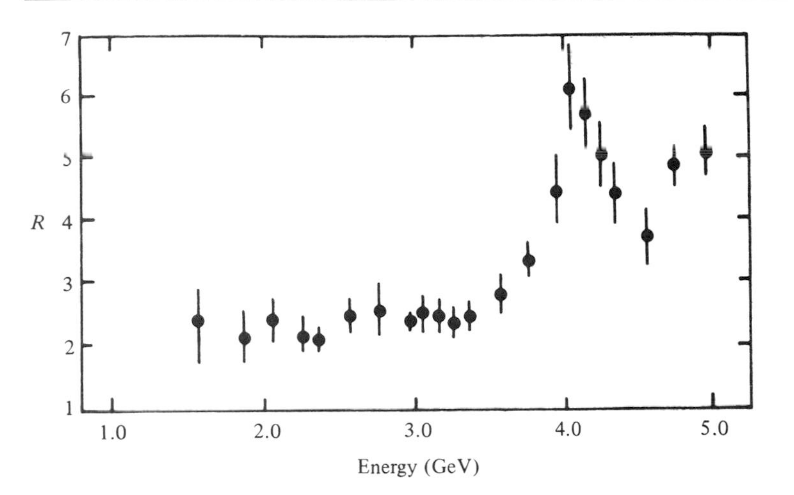

Figure 10.4. Hadron to muon ratio (R) as a function of the total energy of the colliding electron and positron beams. The bars mark the experimental error.

Richter completed his studies at the Massachusetts Institute of Technology (MIT) in 1956. (His brother, Charles, is the geologist who suggested the Richter scale for earthquakes.) He participated in the construction of a pair of experimental storage rings for electrons and positrons at the Stanford Linear Accelerator Center (SLAC). After the success of the trials, Richter proposed to build a much larger ring to enable experiments in the GeV range. Because of lack of funds the proposal was shelved, and in the meantime new storage rings were built in France, as well as at Frascati in Italy and at Novosibirsk in the USSR. Richter, however, did not give up and proposed a less costly design which permitted construction of the ring within SLAC's ordinary budget. In 1970 Richter

storage rings at Frascati in Italy and at the Cambridge laboratories of Harvard. In 1974 a group of physicists at SLAC and the Lawrence-Berkeley Laboratory, headed by Burton Richter and Gerson Goldhaber, initiated a series of experiments aimed at measuring R more precisely and at higher energies, using the SPEAR magnetic detector. In the first experiments it was found that at energies of up to about 3 GeV (1.5 GeV for each beam) R varies between 2 and 2.5, in good agreement with the theory which assumes quarks in three flavours and three colours. Above 3 GeV a significant increase in the hadron to muon ratio was observed. The increase agreed in principle with the forecast of the formation of an additional quark. But instead of R stabilizing at about $3\frac{1}{3}$ as expected, it rose to much higher values and fluctuated in an unexpected manner, as shown in Fig. 10.4. It was suggested that unknown resonances that begin to appear at 3 GeV influenced the results.

In August 1974 Richter decided to re-examine suspicious energy regions and this time to change the energy of the particle beams in much smaller steps (intervals of about 2 MeV instead of 200 MeV as in the first series of experiments). Special steps were taken to make the beam energy sharper and to measure it accurately. In this series of experiments a great surprise awaited the researchers. At a total energy of about 3.1 GeV the detector showed a hundredfold increase in the rate of hadron formation! The stunned physicists present at SPEAR at the time thought there was a fault in the

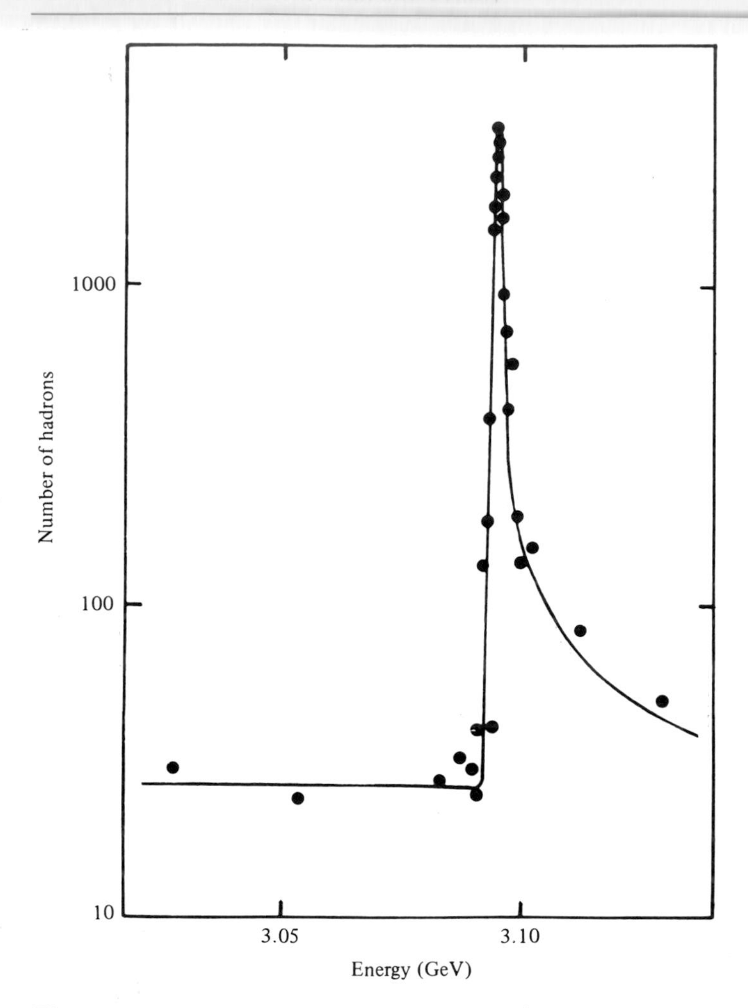

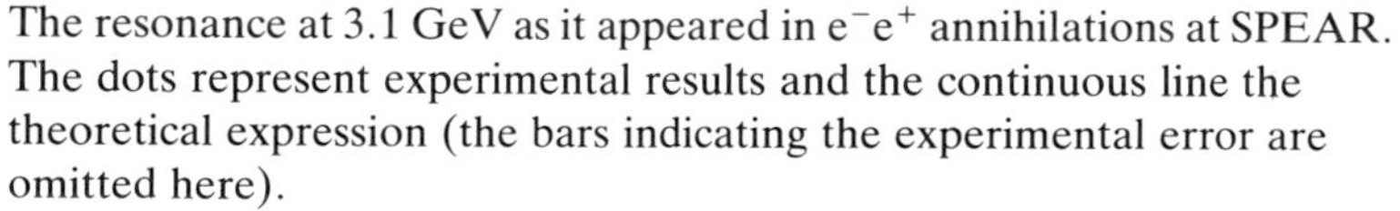

Figure 10.5. The resonance at 3.1 GeV as it appeared in e^-e^+ annihilations at SPEAR. The dots represent experimental results and the continuous line the theoretical expression (the bars indicating the experimental error are omitted here).

(jointly with John Rees) was put at the head of a team for setting up the storage ring, which was named SPEAR. Work was completed in 1973. The budgetary constraints were only too clear from the outward appearance of the machine, which was gifted, according to witnesses, with the grace of the Berlin Wall – bare

instruments, but repeated experiments confirmed the existence of a high, very narrow resonance, shown in Fig. 10.5. Further investigations showed the resonance to be a meson of spin 1. The narrow width of the new resonance (about 0.07 MeV) indicated a lifetime of about 10^{-20} seconds. This was even more surprising, because an ordinary hadron of such a great mass (three times the mass of the proton) would be expected to decay within 10^{-23} seconds via a strong interaction.

As it happened, at that very time the same particle was discovered at another laboratory in the USA, in a completely different type of experiment. At the Brookhaven National Laboratory, in the state of New York, scientists headed by Samuel Ting of MIT were measuring the energy of electron–positron pairs formed in the collision of protons and a beryllium target. They found that this energy was concentrated very sharply around 3.1 GeV. Here too the scientists were quite surprised and the experiments were repeated again and again to make sure that the results were not due to instrumental faults.

concrete and an abundance of wires. But SPEAR vindicated the hopes that had been pinned on it. It made possible the investigation of electron–positron annihilations at energies higher than at any other storage ring at that time – up to 9 GeV – and an astonishing number of important discoveries were made with it. (For the construction of SPEAR, see section 8.4.)

10.3 J, psi, charmonium

On 11 November 1974, Ting visited SLAC and when he met Richter he remarked, 'Burt, I have some interesting physics to tell you about', to which Richter replied, 'Sam, I have some interesting physics to tell *you* about'. That was the first time each had heard of the other's discovery. Comparison of the data showed that both had discovered the same particle – an electrically neutral particle that could be formed both in annihilation of an electron and positron and in interaction between protons and nucleons, and could decay both into an electron–positron pair or into hadrons. In an unforgettable lecture at SLAC the two scientists presented the results of their research before a huge, completely overcome audience that included not only the physicists but also many of the technicians and secretaries of SLAC. Two short papers were sent that week to the Physical Review Letters. Richter's group called the particle psi (ψ), and Ting's group called it J (J resembles the symbol for Ting in Chinese). The name psi is more commonly used today, and we shall keep to it (the notation J/ψ is also common).

The discovery of psi was greeted with great excitement by the physical community. There were those who called the discovery 'the most exciting event since the discovery of the strange particles'. Others simply referred to it as the 'November Revolution'. Soon the particle was observed in other accelerators throughout the world, and its properties were thoroughly investigated. A mere two years after the discovery, Richter and Ting were awarded the 1976 physics Nobel prize.

The theoreticians suggested a plethora of explanations for the new particle. The articles on this topic, which appeared in 1975 like mushrooms after the rain, showed some fertile imaginations at work. The proposed models ranged from conservative ones such as that which viewed the new particle as a bound pair of $\overline{\Omega}^+\Omega^-$ to a model which postulated an exotic particle capable of changing its properties like a chameleon, breaking the laws of conservation of electric charge and parity. But most of these models were soon

Ting came to the USA at the age of 20 and completed his studies at the University of Michigan in 1963. From 1966 he worked at the DESY accelerator in Hamburg, where his group built an advanced sensitive detector system ('double-armed spectrometer') capable of identifying individual electron–positron pairs in a background of millions of hadrons, and measuring their energy. In the summer of 1972 Ting began working at Brookhaven's 33-GeV synchrotron, heading a group of physicists from MIT and Brookhaven. His group set up an improved version of the double-armed spectrometer and began a series of experiments aimed at finding unknown particles in the energy range 1.5–5.5 GeV. For a time they found nothing, but then in August 1974 a sharp resonance around 3.1 GeV was found. By October it was already clear that at that energy a new neutral particle, relatively long-lived, was formed. Five years after this discovery, Ting headed a group of 57 physicists from various countries who found, in the PETRA storage ring at DESY, the 'three-jet' events which are regarded as indirect experimental evidence for the existence of gluons (see section 9.8).

proved false and were discarded. Repeated experiments and theoretical considerations consistently pointed to one explanation – the psi was none other than a meson containing the much talked about charmed quarks, or more precisely it consisted of $\bar{c}c$, like the known mesons ρ^0 and ω^0 consisting of $\bar{u}u$ and $\bar{d}d$ combinations, or even more like the ϕ^0, assumed to consist of $\bar{s}s$.

Since all the quantum numbers of the psi are identical to those of the photon, it can be formed directly from the virtual photon produced in e^+e^- annihilation, without the accompaniment of another particle. The physicists Thomas W. Appelquist and David Politzer of Harvard had predicted the existence of such a particle prior to its discovery and called it charmonium. They even predicted that it could be formed in electron–positron annihilation.

The most important experimental fact that any model of the new particle had to explain was its extremely long lifetime – 1000 times longer than expected for such a heavy particle. On the face of it, the long lifetime of the psi could not be attributed to the conservation of charm, since the net charm of a charmed quark and its anti-quark is zero, and the psi could decay into charmless particles by a strong interaction! However, the flavour of a single quark (or anti-quark) can change only by a weak interaction, and therefore a strong decay of the psi is possible only through annihilation of the charmed quark and anti-quark, and this is a relatively slow process. It had been known for some time that the annihilation of the quarks s and $\bar{s}$ was a surprisingly slow process, which accounted for the relatively long lifetime of the ϕ^0 meson, and the heavier the quark the slower the process.

Identification of the psi as a state of 'charmonium' supplied, therefore, a convincing explanation of its long lifetime. The fact that the psi could decay into leptons or hadrons was also explained by its identification as a charmonium bound state decaying by annihilation, a process in which the original quantum numbers are obliterated. Examination of the decay products showed that the psi had spin 1 and negative parity. This was consistent with the assumption that the two quarks in the psi had parallel spins but no orbital angular momentum relative to each other. From the mass of the psi, the mass of the charmed quark was deduced to be 1.5 GeV approximately.

10.4 The spectrum of charmonium

Having identified the psi as a state of charmonium, physicists were eager to find particles which represented other states of this system. They did not have to wait long. About two weeks after the dramatic

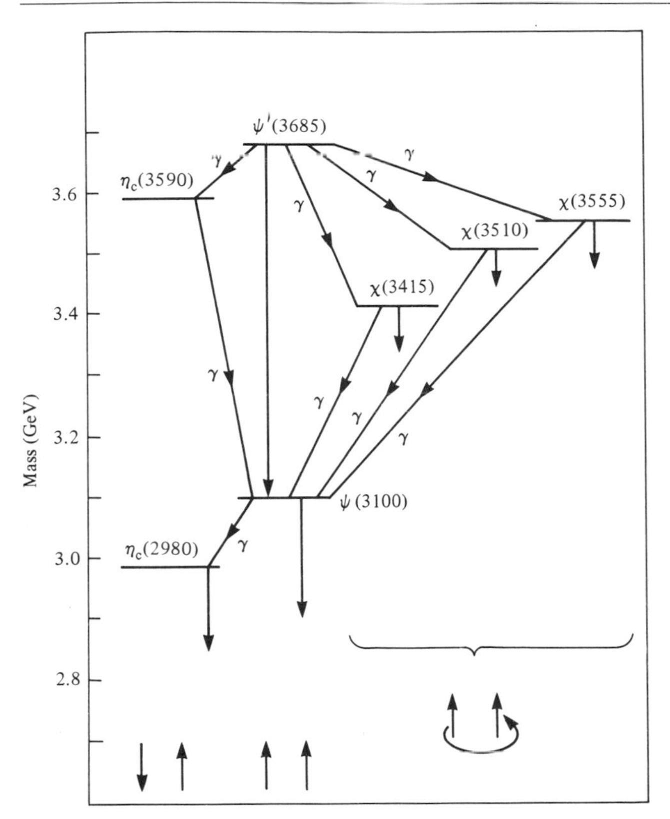

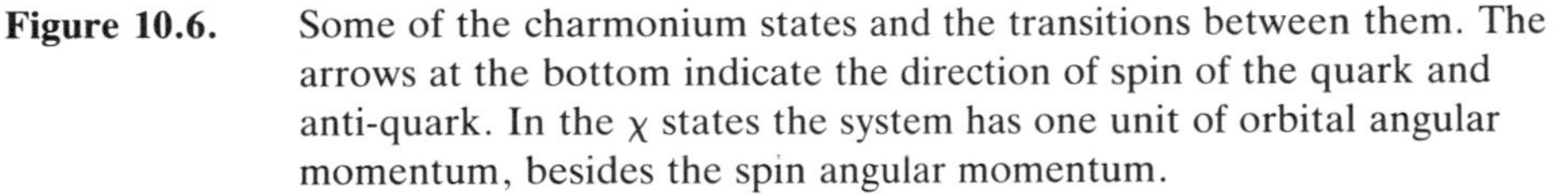

Figure 10.6. Some of the charmonium states and the transitions between them. The arrows at the bottom indicate the direction of spin of the quark and anti-quark. In the χ states the system has one unit of orbital angular momentum, besides the spin angular momentum.

discovery of the psi, Richter's group found another narrow resonance around 3.7 GeV, which decayed in about half the cases to a psi and two pions. It was denoted ψ′. A few weeks later a wide resonance was revealed around 4.1 GeV, and called ψ″. (Later it turned out that around 4.1 GeV there was a group of close resonances.) It was proved that all these particles were excited states of the psi, that is they were composed of the pair $c\bar{c}$ with parallel spins, which were 'less bound' than in the psi. In the course of time more states of charmonium were discovered, and before the eyes of the physicists an almost complete diagram of the charmonium states was unfolded as shown in Fig. 10.6. The states in which c and $\bar{c}$ have parallel spins and zero orbital angular momentum (ψ, ψ′) are sometimes called orthocharmonium. In the states η_c(2980) and η_c(3590) the spins are antiparallel and form a meson of spin zero. Those states are sometimes called paracharmonium. In the three states on the right of the diagram the spins of the quark and anti-quark are parallel or antiparallel and in addition they have one unit of orbital angular momentum, summing up to a total spin of 0, 1 and 2.

The theoreticians pounced on charmonium as on a gold mine. A bound state of a particle and its anti-particle is a relatively simple system to analyse, and the fact that the states of charmonium appeared as narrow resonances in a region free of other resonances simplified the investigation. Fig. 10.6 is reminiscent of the energy levels of an atom or a nucleus. Indeed, in principle, the states of charmonium to a large extent resemble the energy levels of the electron in the hydrogen atom, or the energy states of the deuterium nucleus. (In all three cases the energy levels spring from a change in the state of two particles relative to each other – c and $\bar{c}$ in charmonium, an electron and proton in the hydrogen atom, and a proton and neutron in the deuterium nucleus.)

Just as the study of the energy levels of the hydrogen atom revealed the basic facts about atomic structure, and the investigation into the nuclear energy states elucidated the structure of the nucleus, by fathoming out the charmonium states the physicists learnt that the forces between quarks were also governed by the same fundamental laws of quantum mechanics. Even the methods of research were similar. Just as the hydrogen atom levels were revealed by the photons emitted by the atom, most of the charmonium states were detected by the gamma rays emitted in the transition from one state to another. (These transitions are denoted by γ in Fig. 10.6.) Indeed, only the states of orthocharmonium produce peaks such as the one shown in Fig. 10.5, since only these states have quantum numbers which are identical to those of the photon, and can be formed directly from the virtual photon which appears in the e^+e^- annihilation. The other states have an angular momentum which differs from that of the photon, and therefore they cannot be formed directly from those virtual photons but only from the decay of a ψ or ψ'. These states could therefore be discovered only by detecting the gamma rays emitted in the decay processes.

A particle detector which contributed a lot to exploring the energy states of charmonium at SPEAR was the 'crystal ball'. Designed specifically to detect high-energy photons and measure their energy, this detector consisted of 732 large sodium iodide crystals which surrounded the collision zone of the e^+e^- ring in a sphere-like structure, 2 metres in diameter. Photomultipliers arranged around the sphere of crystals detected the scintillations caused by the passage of high-energy photons in the crystals.

Among the important accomplishments of the crystal ball have been the study of 'glueball' states at 1.44 and 1.67 GeV. The first of these resonances was discovered in 1981 at SPEAR by a detector named Mark II. It stood forth as a prominent peak in the energy distribution of $K_s^0 K^+ \pi^-$ and $K_s^0 K^- \pi^+$ appearing among the decay products of the psi. The crystal ball was employed to investigate this resonance and proved that it had a zero spin and negative parity. This, as well as another particle discovered by the crystal ball at 1.67 GeV (with spin 2), were interpreted as two energy states of the 'glueball' – a short-lived bound system of two gluons having opposite colour pairs.

After operating for three years at SPEAR, the crystal ball was moved to DESY in Hamburg in 1982, and operated by an international group of Americans and Europeans. This is one of the signs of the international collaboration epoch in science. A device planned and constructed in one country is transferred to another country, to be operated by an international team.

10.5 The quest for charm

The discovery of the various states of charmonium, whose masses had in some cases been predicted by the theoreticians with rather good accuracy, convinced even the sceptics among the physicists that the psi and its hangers-on consisted of a new quark and anti-quark pair. But to prove that the new quark had this special flavour called 'charm', it was necessary to find a particle with non-zero charm, such as a meson composed of $c\bar{u}$ or $c\bar{d}$, in which, unlike in the psi, the charms of the two constituents would not cancel each other out.

Thus, at several accelerators the hunt for charm was on. It was anticipated that charmed particles would have several identifying features: they would be produced in pairs in strong or electromagnetic interactions and decay in weak interactions like the strange particles. They would have relatively long lifetimes, though not as long as the strange particles (because of their higher masses). The decay of the charmed quark was believed to be preferentially into an s quark, rather than the lighter u or d quarks, which meant that among the decay products of a charged particle there would usually be one strange particle. The physicists expected to find leptons as well among the decay products (typically of a weak interaction) and there were also certain assumptions concerning the masses of the lightest charmed mesons and baryons.

Despite all this information, the search for the charmed particles was unexpectedly long and arduous. Their lifetime of about 10^{-13} seconds was too short to give a detectable path in a bubble chamber;

even when moving at close to the speed of light, the particle would traverse only a few hundredths of a millimetre before disintegrating. On the other hand, this lifetime was seven orders of magnitude longer than the lifetime of the psi, and because of the relation between lifetime and resonance width, the resonance of a charmed particle would be ten million times narrower than the already narrow resonance of the psi (shown in Fig. 10.5). The energy of the beam in an accelerator cannot be changed in such small steps and therefore such particles cannot be detected in the same way as the psi was detected in the storage ring. There is, however, an alternative way to discover short-lived particles, by measuring the energies of their expected decay products. It was by this method (described in section 8.1 and Figs. 8.3–8.6) that Ting's group had made their independent discovery of the psi. They had found that most of the electron–positron pairs formed in the collision between protons and a target had energies of around 3.1 GeV, and concluded that they must be the products of a short-lived intermediate particle with that mass.

But even the efforts to find charmed particles by their decay products failed at first. It was found that in the energy range where charmed particles were expected to be produced, a prodigious number of conventional particles were formed and looking for the decay products of the carriers of charm was like looking for a needle in a haystack. Indeed, in the years 1975–76, results were obtained in several accelerators that could be interpreted as the footprints of charmed particles (in the bubble chambers at Brookhaven and CERN, interactions between neutrinos and nucleons were being investigated, and the photographs showed several events which presumably bore the sign of charm, while at Fermilab similar interactions provided two-lepton events which were deemed to spring from a charmed particle with a mass of around 2 GeV). But these findings – some of which were announced by special press conferences and made headlines in the daily newspapers – did not supply irrefutable evidence of a charmed particle and did not permit exact determination of its mass.

In mid-1976, the storage rings made history once more. A group of 41 scientists from SLAC and Berkeley (many of whom had participated in the discovery of the psi) detected – with the aid of SPEAR – the lightest charmed mesons, and succeeded in measuring their masses with great accuracy. The group was headed by Gerson Goldhaber and Francois Pierre. With the aid of the versatile magnetic detector, the group was studying electron–positron annihilations whose products contained kaons and pions of total charge zero ($K^+\pi^-$ or $K^-\pi^+$. It will be recalled that one of the

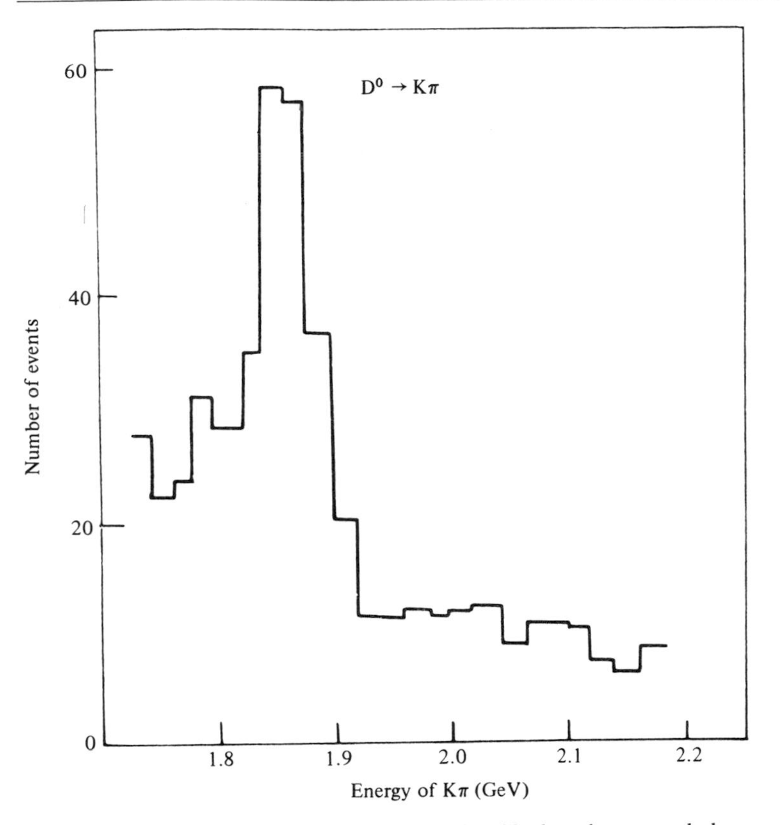

Figure 10.7. The resonance at 1.86 GeV, which was identified as the neutral charmed meson, D^0.

characteristics of a charmed particle is the presence of a strange particle among its decay products). The data gathering lasted many months, but in the spring of 1976 it was established beyond doubt that when the sum of the energies of the electron and positron exceeded 3.73 GeV, the energies of the kaon and pion produced in the annihilation formed a prominent peak around 1.865 GeV (see Fig. 10.7). This resonance was identified as a neutral charmed meson, denoted D^0, composed apparently of the quarks $c\bar{u}$. The fact that the resonance appeared only at collision energies greater than 3.73 GeV is easily explained as follows: the annihilation is an electromagnetic interaction in which charm is conserved, and therefore if a particle of charm +1 is formed, a particle of charm −1 must be formed as well; so the collision energy must be at least double the mass of the lightest charged meson. The formation of charmed particles in pairs is illustrated in Fig. 10.8, which is a computer reconstruction of an event interpreted as the formation of ψ'' in an e^+e^- annihilation, and its decay into $D^0\bar{D}^0$.

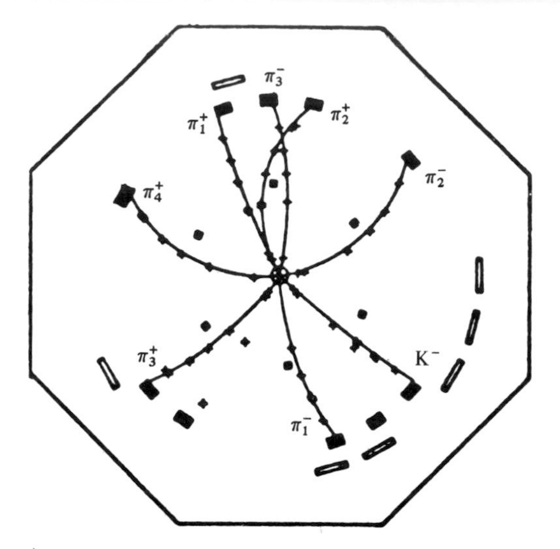

Figure 10.8. A computer reconstruction of an event which was recorded by the magnetic detector at SPEAR, and interpreted as the formation and decay of a $D^0\bar{D}^0$ pair, through the following processes:

$$e^+ + e^- \rightarrow \psi'' \rightarrow D^0 + \bar{D}^0$$
$$D^0 \rightarrow K^- + \pi_1^+$$
$$\bar{D}^0 \rightarrow K_s^0 + \pi_2^+ + \pi_3^+ + \pi_1^- + \pi_2^-$$
$$K_s^0 \rightarrow \pi_4^+ + \pi_3^-$$

By a method similar to that used to discover the D^0, the charged charmed mesons D^+ and D^- (composed of $c\bar{d}$ and $\bar{c}d$) were detected at SLAC. It was found that each had a mass of 1.869 GeV. Later, particles possessing both charm and strangeness were discovered: the mesons F^+ and its anti-particle F^-, the composition of which are apparently $c\bar{s}$ and $\bar{c}s$ respectively. In addition, a charmed baryon was found, named the charmed lambda (Λ_c), presumably built of the quarks udc. Due to their weak decay, the lifetimes of the charmed particles are relatively long, compared to non-charmed particles of the same mass, and are of the order of 10^{-13} seconds.

At the end of the 1970s, photographs of the formation and decay of charmed particles were at last obtained, using a sensitive photographic emulsion which, after being developed, was scanned under the microscope. One such photograph is shown in Fig. 10.9. A proton of energy 80 GeV hit the photographic plate and formed a 'star' of particles. One of these was a neutral particle which decayed at a distance of about 0.1 mm from the centre of the star, at the position marked with a circle. Examination of its decay products, marked 1d–4d showed that the original particle was a $\bar{D}^0$. Its mass

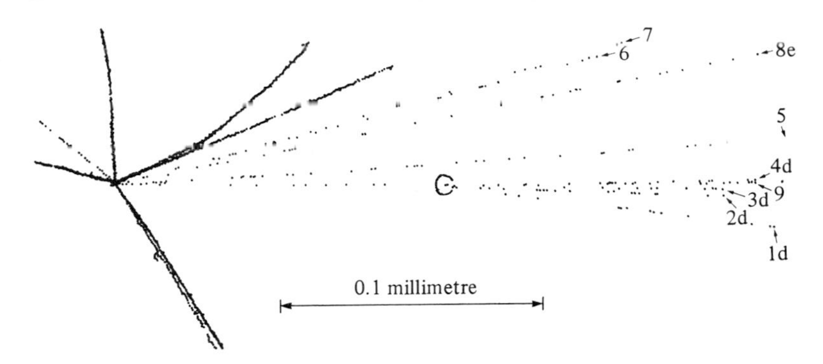

Figure 10.9. A 'star' produced by the hit of a high-energy photon on a photographic plate. The tracks 1d–4d apparently correspond to the decay products of a $\bar{D}^0$ which emerged from the centre of the star and decayed in the region marked by a circle.

according to this experiment was 1866 ± 8 MeV. The experiment was carried out at CERN, with the participation of physicists from several European countries (the paper that appeared in *Physics Letters* on 28 January 1980 was signed by no fewer than 96 physicists!).

10.6 Another quark enters the arena

With the existence of charm established, the list of quarks comprised four types or 'flavours'; some of their properties are summarized in Table 10.1.

The protons and neutrons in atomic nuclei (that is, all the stable hadronic matter in our world) are composed of the light quarks u and d, while the heavier quarks s and c appear in short-lived particles, formed in accelerators or in reactions within stars and in cosmic rays.

The question which now arose was: was this the complete list of quarks or would we be able to find particles composed of still heavier quarks by increasing the energy of our accelerators? The answer was given in the summer of 1977 when a group of physicists

Table 10.1. *The main properties of the quarks d, u, s, c*

Quark	Electric charge	Baryon number	Strangeness	Charm
d	$-\frac{1}{3}$	$\frac{1}{3}$	0	0
u	$\frac{2}{3}$	$\frac{1}{3}$	0	0
s	$-\frac{1}{3}$	$\frac{1}{3}$	-1	0
c	$\frac{2}{3}$	$\frac{1}{3}$	0	1

Lederman's group began its research at Brookhaven in 1968, and in 1975 transferred to Fermilab. The task it set itself was not easy. When a beam of protons of energy 20 or 30 GeV hits matter, a great many hadrons are formed, and only a few muons. In order to isolate the handful of muons from among the profusion of other particles, the detectors were placed behind an iron wall 3 metres thick. This wall easily stopped all the hadrons, while allowing the muons, with their much higher penetrating power, to pass through. But this arrangement had one big drawback: the energy of the muons changed in the course of their collisions with the iron atoms, so that the values measured by the detectors were not the original energies of the muons. The investigators hoped, however, that the distortion would be bearable, and that any prominent peak in the original energy distribution of the muons would show up in the recorded results as well. Alas, they were cruelly refuted! They did find a blurred bump around 3 GeV, but could not establish its nature. Later, when Richter and Ting discovered the psi, Lederman's group realized that the blurred bump was nothing but the remains of the sharp resonance of the psi, after distortion by their system. Because of the clumsiness of their apparatus, the discovery of the psi had been denied them, and with it the Nobel prize. But they were not discouraged, and did not abandon the

at Fermilab, headed by Leon M. Lederman, announced the discovery of a new meson which was much more massive than any known particle. The group, which for more than 9 years had been investigating the energies of muon pairs (μ^+, μ^-) formed when high-energy proton beams hit a beryllium target, found a narrow unknown resonance at 9.46 GeV. It was clear that the heavy particle could not be composed of the known quarks but must be made up of a new type of quark. The new particle was denoted by the Greek letter upsilon Υ (sometimes also written Y). Additional resonances regarded as excited states of the upsilon were later found at 10.02 and 10.35 GeV, and denoted Υ′ and Υ″, respectively.

About a year after its discovery at Fermilab, the upsilon and its excited states were seen also in electron–positron annihilations in the storage ring DORIS in Hamburg. It was proved to be a meson, and since such a heavy meson could clearly not be composed of the known quarks, this was taken as proof of the existence of a fifth quark, heavier than all the known ones.

The fifth quark was denoted by the letter b, and the upsilon was taken to be composed of a bound pair $b\bar{b}$, in analogy with the psi which was identified as the bound pair $c\bar{c}$. From a study of the upsilon and its decay products, it was concluded that the electric charge of the b quark was $-\frac{1}{3}$, and from the narrow width of the resonance (which, it will be remembered, points to the particularly long life of the particle) it was concluded that the fifth quark, like the s and c, carries a unique 'charge' of its own which is conserved in strong and electromagnetic interactions but not in weak ones. This charge, or flavour, was called 'beauty' and denoted B. On the assumption that the mass of the upsilon is about equal to the sum of the masses of the quark and anti-quark of which it is composed, the mass of the b quark comes out to be about 4.7 GeV, which is three times greater than the mass of the c quark.

The detection of the upsilon spurred an ardent search for mesons composed of a b (or $\bar{b}$) and a light quark (u or d), similar to the chase after charmed mesons following the discovery of the psi. The theory predicted two charged mesons ($B^+ = \bar{b}u$, $B^- = b\bar{u}$) and two neutral ones ($B^0 = b\bar{d}$, $\bar{B}^0 = \bar{b}d$), and could describe their expected properties.

In this project the pre-eminence of storage rings over fixed-target accelerators was shown once more. Actually the entire research was done in one storage ring which was inaugurated just in time and with exactly the right energy, at Cornell University, in the pastoral town of Ithaca in the state of New York. The Cornell electron–positron storage ring (CESR) began operation in 1979, with a maximal

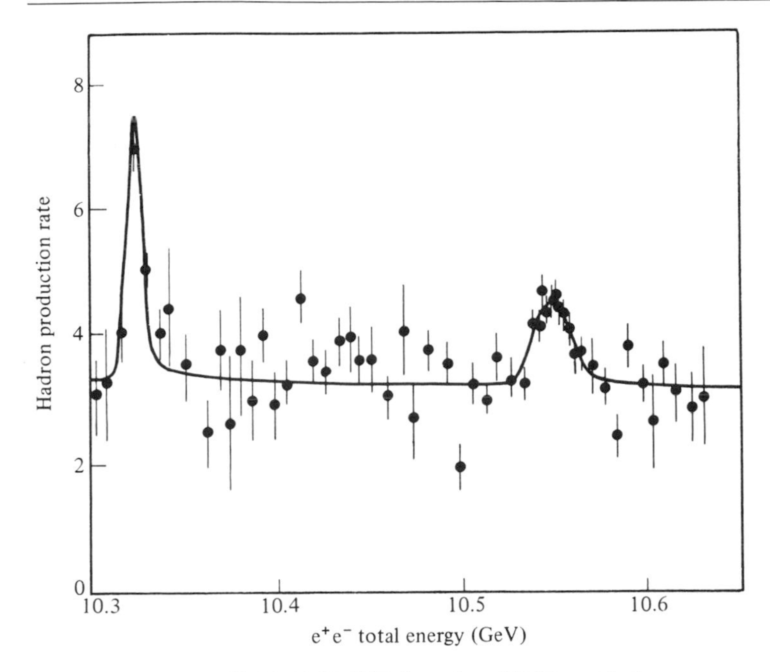

Figure 10.10. The resonance at 10.58 GeV (Υ‴) found at CESR, with the narrower, previously known, resonance at 10.35 GeV (Υ″).

battlefield. Instead, they set about improving their system, so that it could measure energies more accurately. From the success of Ting and his group, who had accurately measured the energy of e^+e^- pairs amid a large preponderance of hadrons, they learnt that the mission could be done. And indeed, after replacing the iron wall with beryllium (which has lighter atoms than iron and thus has a smaller effect on particles passing through it), and after improving the detectors and employing sophisticated computer programs to reconstruct the original muon energies from the measured energies, Lederman and his group succeeded in seeing the psi as a clear prominent peak at 3.1 GeV. They

centre-of-mass energy of 16 GeV. Early in 1980 it provided an important finding: a new resonance at 10.58 GeV, which was deemed to be another excited state of the upsilon. Unlike the lower energy states, which were very narrow, this new resonance was quite wide, indicating a much faster decay (see Fig. 10.10). It was immediately suspected that the shorter life span of this upsilon state (designated Υ‴) stemmed from its being above the threshold for decaying into a pair of B and $\overline{B}$ mesons, a decay which requires neither alteration of quark flavours (a weak interaction) nor the annihilation of a quark and anti-quark (a suppressed strong interaction). Strong indirect evidence for this explanation was a dramatic increase in the yield of high-energy electrons and muons observed when the total energy of the electron and positron beams was close to 10.58 GeV. These energetic leptons were difficult to account for unless they were decay products of the B mesons (which can decay only by weak interactions owing to the conservation of beauty in strong interactions). Another propitious omen was the remarkable number of charged kaons. It was believed that B mesons decayed predominantly to charmed D mesons whose subsequent decays produced K^+ and K^-.

All the evidence implied that if one could tune the e^+e^- total then proceeded to higher energies, and at about 9.5 GeV they found at last what they had been seeking for almost 10 years: a new resonance that nobody had seen before.

In 1982 Lederman was awarded the Wolf prize, together with M. Perl (see below).

energy to 10.58 GeV, one will find B mesons among the decay products of the Υ'''. However, to search for B events among all the other decays was not an easy task. Even when a B meson is produced, it has a great variety of decay modes. Some of these decays yield numerous particles including neutral particles that escape detection. Thus, the combinatorial analysis involved in identifying the B mesons and their intermediate decay products might be unimaginably complicated. Luckily for the CESR people, they enjoyed an almost entire monopoly in the energy range relevant for producing the B mesons and could work out their problems sedately without the fear of being overtaken by another laboratory. The DORIS storage ring (at DESY, Hamburg) had been stretched to its energy limits in order to observe the Υ' at 10.0 GeV, but neither this collider nor SPEAR at SLAC could at that time reach 10.58 GeV. Ironically, the new generation of storage rings at SLAC (PEP) and DESY (PETRA) with centre-of-mass energies of 38 GeV were quite ineffective at low energies. Later, the VEPP-4 e^+e^- storage ring at Novosibirsk in central Siberia joined the race with the convenient total energy of 14 GeV (and DORIS was upgraded to 11 GeV) but it was already too late to overcome CESR's advantage.

The role of searching for B mesons in the reaction $e^+e^- \to B\bar{B}$ was mainly carried out on the CLEO detector – an array of detectors planned and operated by a collaboration of 79 physicists from eight institutions. The collaboration decided to restrict the search to a few specific B decays which could be readily analysed, such as

$$B^- \to D^0 + \pi^-$$
$$D^0 \to K^- + \pi^+$$

and its charge conjugate for B^+ decay, or, such as

$$B^0 \to \bar{D}^0 + \pi^+ + \pi^-$$
$$\bar{D}^0 \to K^+ + \pi^-$$

The theory predicted that the negatively charged b quark in the B^- meson would decay preferentially to the positively charged c quark (by emitting a virtual W^-) and that the c quark would readily decay into the negative s quark (by emitting a virtual W^+) as shown in Fig. 10.11.

During the months of the experiment, each favourable event recorded by the CLEO detector was analysed by the computer in order to evaluate the mass of the putative B meson. Due to the rarity of these events it took several months before enough statistics could be gathered and analysed. By the end of 1982, however, 18 events

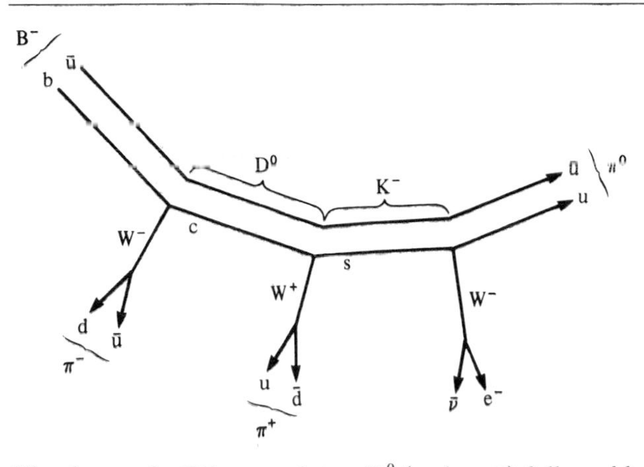

Figure 10.11. The decay of a B^- meson into a D^0 (and a π^-) followed by the decay of the D^0 into a K^- (and a π^+) in terms of the quark model.

provided successful fits to the considered decay sequence. These 18 events were selected from 140 000 e^+e^- collisions. Nine of them corresponded to charged B mesons and nine to the neutral Bs. The mass values computed from these events were found to cluster nicely around 5.274 GeV for both the charged and neutral B mesons. The lifetimes of the B mesons were estimated to be of the order of 10^{-13} seconds.

Thus the list of mesons which are stable against all but weak decay that had included only the π^+, K^+, K^0, D^+, D^0, F^+ and their anti-particles was extended to include B^+, B^0 and their anti-particles. The relative stability of all these particles demonstrates the conservation of the various quark flavours in strong and electromagnetic interactions. The discovery of the B mesons provided therefore a strong confirmation of the existence of a fifth quark flavour.

H. Harari of the Weizmann Institute had originally used the letters b and t for 'bottom' and 'top' in a hypothesis involving a third pair of quarks, back in the early 1970s. This conjecture had been advanced to explain some experimental results that later proved to be erroneous. Harari's nomenclature was nevertheless revived when the third 'generation' of quarks eventually became reality.

Adding the beauty-carrying quark to the quark table brings their number to five. These five quarks can be arranged in two pairs – (d, u) and (s, c), each consisting of one quark with electric charge $-\frac{1}{3}$ and one with charge $\frac{2}{3}$ – and a single quark with charge $-\frac{1}{3}$. There was immediate speculation that the fifth quark also had a partner – a sixth quark, with electric charge $\frac{2}{3}$ and a new kind of charge (or flavour) unique to itself, which was called 'truth'. This quark was therefore named 't'. (In another common notation b and t stand for *bottom* and *top*, but we prefer beauty and truth here, as they seem to be more attractive descriptions.) The search for particles containing the sixth quark appears to have come to a happy conclusion in 1985, when first signs of a meson containing the truth-carrying quark were detected at CERN (see section 10.8).

Table 10.2. *The most important properties of the quarks d, u, s,c, b and t*

	Q	A	S	C	B	T
d	$-\frac{1}{3}$	$\frac{1}{3}$	0	0	0	0
u	$\frac{2}{3}$	$\frac{1}{3}$	0	0	0	0
s	$-\frac{1}{3}$	$\frac{1}{3}$	-1	0	0	0
c	$\frac{2}{3}$	$\frac{1}{3}$	0	1	0	0
b	$-\frac{1}{3}$	$\frac{1}{3}$	0	0	1	0
t	$\frac{2}{3}$	$\frac{1}{3}$	0	0	0	1

Q denotes electric charge, A – baryon number, S – strangeness, C – charm, B – beauty (or bottom), T – truth (or top).

To sum up, we can say that in order to explain the structure of all the hadrons in terms of the quark theory we need six different quarks (each havings its own anti-quark). Each quark can, according to quantum chromodynamics, appear in three colours. The forces between the quarks in a hadron are carried by gluons, of which there are eight types. All the new hadrons discovered fit well into the assumption that the possible combinations of quarks and anti-quarks in hadrons are: three quarks (forming a baryon), three anti-quarks (forming an anti-baryon) and a quark–anti-quark pair (forming a meson). The most important properties of the quarks d, u, s, c, b and t are summarized in Table 10.2.

10.7 The number of leptons grows too

Up to 1975 only four leptons (and their anti-particles) were known. These were the electrons (e^-), the muon (μ^-) and their two neutrinos (ν_e, ν_μ), along with their anti-particles: the positron (e^+), the positive muon (μ^+) and their two anti-neutrinos ($\bar{\nu}_e$, $\bar{\nu}_\mu$).

In 1975 Martin Perl and his group, working at Stanford, discovered another charged lepton, which they called tau (τ). The discovery was important not only because no new lepton had been found for many years, but also because of the large mass of the new lepton – approximately double that of the proton – which made the name 'lepton' (meaning 'light' in Greek) somewhat inappropriate.

Because of its short lifetime (of the order of 10^{-13} seconds) it is very difficult to 'see' the tau directly, and it was discovered and studied by methods used for detecting resonances. Perl and his group worked with the SPEAR storage ring at SLAC, and used the all-powerful magnetic detector to study electron–positron annihilations which yielded as end products an electron and muon:

$$e^+ + e^- \rightarrow e^+ + \mu^- + \ldots \text{ or } e^+ + e^- \rightarrow e^- + \mu^+ + \ldots$$

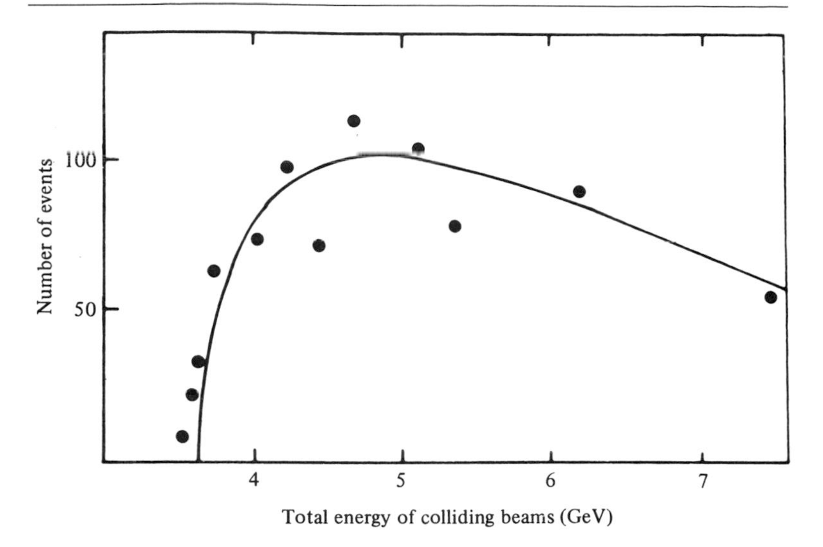

Figure 10.12. The production rate of electron–muon pairs as a function of the total energy of the colliding e^-e^+ beams. The dots mark experimental results and the continuous line, the best fit of the theoretical expression. The threshold energy above which electron–muon pairs begin to appear is 3.6 GeV. The moderate slope of the high-energy side of the graph indicates that the short-lived particle is a lepton, since in hadronic resonances the drop should be much faster.

The dots in the equations indicate that additional particles not recorded by the detectors are formed, as required by the laws of conservation of the lepton numbers (see section 7.6).

The researchers found that the formation of an electron–muon pair in the annihilations was rather rare. They also found that such pairs appeared only when the overall energies of the colliding beams exceeded 3.6 GeV (see Fig. 10.12). They concluded that initially the annihilation produced a particle–anti-particle pair, in which each member was short-lived and had a mass of around 1.8 GeV; in the decay of these particles the electron and muon were formed. This conclusion was ratified by careful study of the recorded reactions. Various pieces of experimental evidence (e.g. the shape of the graph in Fig. 10.12 which represents the probability of formation of an electron–muon pair as a function of energy) indicated that this new particle was a lepton. It was the third charged lepton known – after the electron and muon – and so it was named the tau (τ), standing for the word '*triton*' meaning 'third' in Greek. Like the electron and muon, the tau could carry a positive or a negative electric charge, and it was assumed that it had its own neutrino ν_τ. It was thought that in an e^+e^- annihilation producing $e^+\mu^-$, the reactions were actually as follows:

$$e^+ + e^- \rightarrow \tau^+ + \tau^-$$
$$\tau^- \rightarrow \mu^- + \bar{\nu}_\mu + \nu_\tau$$
$$\tau^+ \rightarrow e^+ + \nu_e + \bar{\nu}_\tau$$

The detectors of course could not detect neutrinos, and recorded only the positron and muon. Indeed, calculations of the momentum and energy of the electron and muon in each of the recorded events confirmed that each of these particles was accompanied by two neutral particles which carried off part of the energy and momentum of the tau. Fig. 10.13 depicts schematically the process of formation of a $\tau^+\tau^-$ pair in a storage ring and their decay to a positron, negative muon and four neutrinos. (The circle represents a cross-section of the storage ring; the colliding beams are perpendicular to the plane of the paper.) The disintegration of taus can, of course, also yield an e^-e^+ or $\mu^-\mu^+$ pair, but since in these cases one cannot tell whether the products were formed directly in the annihilation or were the decay products of an intermediary heavy lepton, the study had to be limited to the rarer events of muon–electron formation. Because of the rareness of these events it took many months of intensive work on the part of Martin Perl and his group before they had collected a sufficient number of events to be able to say with certainty that a new lepton had indeed been discovered. Perl was awarded the Wolf prize in Jerusalem in 1982.

The tau was further investigated at the larger storage rings PETRA (at DESY, Hamburg) and PEP (at SLAC, Stanford). Its mass was determined to be 1.78 GeV, and its lifetime, about 3.5×10^{-13} seconds.

With the discovery of the tau, and assuming that it really has its own type of neutrino, the number of known leptons stands at six particles and six anti-particles, which can be arranged in three 'families' as shown in Table 10.3.

Table 10.3. *Families of leptons*

	Particle	Anti-particle	Mass (MeV)	Lifetime (seconds)
Electron family	e^-	e^+	0.51	Stable
	ν_e	$\bar{\nu}_e$	0?	Stable
Muon family	μ^-	μ^+	106	2.2×10^{-6}
	ν_μ	$\bar{\nu}_\mu$	0?	Stable
Tau family	τ^-	τ^+	1784	3.5×10^{-13}
	ν_τ	$\bar{\nu}_\tau$	0?	Stable?

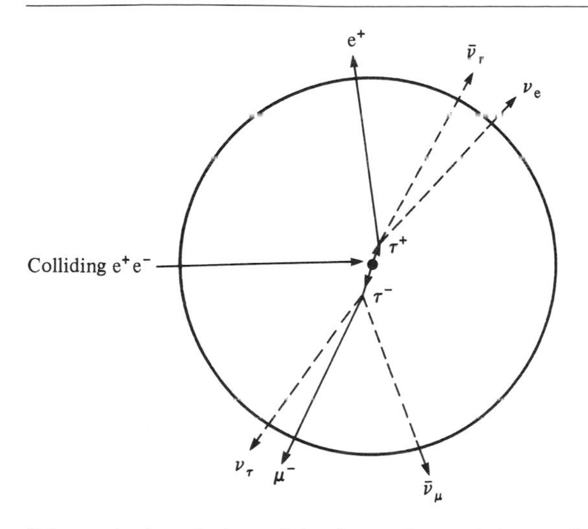

Figure 10.13. Schematic description of the formation and decay of the heavy leptons, τ^+, τ^- in the annihilation of an electron and positron.

This is the place for making an important remark concerning the mass of the neutrino. All the experimental and theoretical data gathered up until the late 1970s were consistent with the hypothesis that the mass of the electronic neutrino and muonic neutrino is actually zero. However, in the years 1979–80 several researchers (among them Reines, who in the 1950s co-discovered the electronic neutrino with Cowan) published experimental results which indicated that the mass of the neutrinos might be non-vanishing, although very small compared to other particles (of the order of 10 eV, or 1/50 000 electron masses). An interesting consequence was that a beam of neutrinos might oscillate between various states, for example ν_e and ν_μ, in a similar way to the behaviour of the neutral kaons (section 7.7) by violating the conservation of lepton numbers. The results, however, were not conclusive, and have not been confirmed by other experiments; thus the question of the neutrino mass is still open to date. In any case, even if the masses of the ν_e and ν_μ eventually turn out to be non-zero, they are certainly very tiny so that one can go on considering them as nil in most of the calculations involved in the analysis of particle reactions.

Our knowledge concerning the tau neutrino is quite meagre, and the hypothesis that it is stable and has a zero (or very small) mass is based on the assumption that it is similar in these respects to the previously observed types of neutrinos rather than on clear experimental results.

We do not know the reason for the immense differences between the masses of the various charged leptons. (The mass of the tau is 18 times larger than the mass of the muon, and about 3500 times as large as that of the electron.) We also do not know whether the discovery of the tau has completed the list of leptons, or whether there are other, heavier leptons yet to be discovered. According to some theories, there is a direct correspondence between the number of quarks and leptons, and thus the discovery of new leptons may imply the existence of additional quarks, and vice versa.

10.8 The discovery of the carriers of the weak force

The Weinberg–Salam–Glashow theory which unified the electromagnetic and weak interactions (section 6.1) has received much experimental confirmation, the most significant of which was perhaps the neutral weak currents, discovered in 1973 (section 10.1), and mentioned in the Nobel Committee arguments in 1979. Further support came from electron accelerators and storage rings at SLAC and DESY. For example, electron-scattering experiments at SLAC in 1978 clearly demonstrated interference effects between the electric and weak forces, as predicted by the theory.

However, the crucial test of the theory, the search for the particles that mediate the weak force, could not be done during the 1970s because of the lack of accelerators of adequate energy. The theory envisaged three such mediator particles and also predicted their masses: the W^+ and W^- of about 80 GeV and the 90-GeV Z^0. They are collectively referred to as 'the weak intermediate vector bosons' (i.e. bosons of spin 1) or, more simply W and Z bosons. It was anticipated that these particles would be produced in abundance when the required energies were attained, and decay in about 10^{-20} seconds in such a way that their existence could be verified and their properties studied.

The most powerful accelerators in the 1970s were the 500-GeV proton synchrotrons at Fermilab and CERN (inaugurated in 1972 and 1976 respectively) but only about six per cent of that energy could be converted into mass in collisions with a stationary target (section 8.4). The largest e^+e^- collider of the 1970s was PETRA (inaugurated at DESY, Hamburg, in 1978) with a centre-of-mass energy of 38 GeV. It was clear that the search for the W and Z bosons called for new accelerators, preferably of the collider type.

Since no known conservation law prohibited the creation of a single Z^0 in e^+e^- annihilations it was expected that an e^+e^- collider with a total energy of 90–100 GeV would suffice for the discovery of the neutral carrier of the weak force, provided that its mass was as anticipated. Such annihilations, however, cannot produce isolated W^+ or W^- because of charge conservation. Thus a 160–170 GeV e^+e^- collider would be needed to produce W^+W^- pairs.

The reckoning is different when a $p\bar{p}$ collider is discussed. The weak force carriers were expected to be produced in a $p\bar{p}$ collision, due to an interaction between a single quark in the proton and a single anti-quark in the anti-proton. For example:

$$u+\bar{d}\rightarrow W^+,\quad d+\bar{u}\rightarrow W^-,\quad u+\bar{u}\rightarrow Z^0,\quad u+\bar{u}\rightarrow W^+ + W^-$$

For the quark and anti-quark to have enough energy for such reactions, the centre-of-mass energy of the proton and anti-proton should be much higher, because each quark carries only part of the energy of the proton. The minimal centre-of-mass energy required to see W^+, W^- and Z^0 in $p\bar{p}$ head-on collisions was calculated to be about 500 GeV. The production rate of the coveted particles was expected to be quite low at 500 GeV but to rise sharply above 1000 GeV.

Apart from the lower energy needed, and the fact that e^+e^- storage rings constituted by the mid 1970s a well-tested technique while $p\bar{p}$ colliders were a rather new and untested device, there was another reason to prefer e^+e^- colliders. Electron and positron beams of the right energies were expected to provide very 'clean' events. That is because a single collision at the appropriate energy may produce a single Z^0, or a pair of W^+W^-. On the other hand, if a Z^0 is for instance produced in a $p\bar{p}$ collision, due to an interaction between a quark and an anti-quark, it is apt to be accompanied by a profusion of other particles (mostly hadrons). These particles would stem from interactions between the other four quarks and anti-quarks in the colliding proton and anti-proton, and would complicate the analysis. Counter to these arguments stood the fact that energy dissipation due to synchrotron radiation is much more severe for electrons than for protons (see section 4.2). Some physicists thought of a third alternative – proton–proton collisions in interlaced rings – as the best way to look for the intermediate vector bosons. But, in 1976 a new daring idea was proposed by physicists from both sides of the Atlantic, which turned the tables and caused the $p\bar{p}$ collider to be the first to get the first look at the weak-force quanta.

The Italian physicist Carlo Rubbia from CERN and the Americans David Cline and Peter McIntyre suggested that instead of building an entirely new accelerator, it would be much cheaper and faster to take an existing proton synchrotron and convert it into a $p\bar{p}$ storage ring. After a thorough analysis the project was approved both at Fermilab and at CERN, and a friendly race began between the two laboratories. Everyone was aware of the fact that the first machine to be completed might well make one of the most significant discoveries of the decade.

The people of CERN decided to go for the minimal energy that would do the job according to calculations. They planned to turn their Super Proton Synchrotron (SPS) into a $p\bar{p}$ storage ring of a total energy of 540 GeV (270 GeV per beam). The plans at Fermilab were more ambitious. They planned to install new superconducting magnets* in the tunnel of the 500 GeV proton synchrotron, then boost the beam energy up to 1000 GeV, and afterwards convert the fixed-target accelerator into a $p\bar{p}$ storage ring of a total energy of 2000 GeV. The design luminosity (a measure of the density of particles in the beam and of the collisions rate) was 10^{31} per square centimetre per second, 10 times more than that at CERN.

At CERN things went smoothly and fast and in 1981 the first $p\bar{p}$ collisions began to take place, while the Fermilab people still struggled with their fixed-target 'Tevatron' (1 TeV = 1000 GeV). In the physics community it is not customary to cheer your favourite team and wish its rivals bad luck, yet we may assume that more than one American physicist quietly prayed that the energy and luminosity of the CERN collider would prove not high enough for a safe identification of the massive sought-after particles, and that it would be the Fermilab machine, with its 2000 GeV and higher luminosity, which eventually would make the great discovery.

A mite of the ambivalent feelings of the American scientists facing the success of the Europeans is revealed in an article written by R. Wilson, former director of Fermilab. Wilson wrote (*Physics Today*, Nov. 1981, p. 99): 'CERN had the daring, and wisdom and expertise, *and* funds to proceed forthrightly, and by a remarkable tour de force has now brought that nightmarishly complicated system into operation. They have observed proton–antiproton collisions. The pain for us in watching this bravura performance is that they have brought to this accomplishment verve and audacity and skill – virtues we thought to be exclusively American. May they reach meaningful luminosity and may they find the elusive intermediate boson. We will exult with them when they do.' Later in the same article Wilson asks himself: 'if the preeminence we have enjoyed for almost half a century is coming to an end'. He admits that although this talk about pre-eminence and about winning races is contrary to the spirit of physics, there still does remain 'a chauvinistic problem, a duality in thinking' which one cannot completely avoid.

* Some metals exhibit zero electrical resistance, or become superconductors when cooled to very low temperatures. An electromagnet made of this metal, when cooled to the appropriate temperature, can produce high magnetic fields without the energy losses associated with ohmic resistance.

In the course of preliminary experiments at CERN, beginning in July 1981, the luminosity was quite low (2×10^{27} per square centimetre per second at the most) and no meaningful events could be detected. Then came several months of adjusting and improving the equipment, and then a second run of 30 days which terminated in December 1982. In this run, the luminosity was raised to 2×10^{28} per square centimetre per second. This was still much less than the designed value of 10^{30}; nevertheless, the computer program aimed at 'hunting' likely events was operated on-line. The number of tell-tale events was very small, but they were crystal clear. In January 1983, the two groups of investigators working independently with two huge detectors (prosaically named UA1 and UA2) gave the first reports on their results in an international conference at the New York yearly meeting of the American Physical Society. Out of 10^9 $p\bar{p}$ collisions examined by each detector, 10^6 were recorded. Of this multitude, only nine events were selected by the computer, but each of these bore the signature of one of the Ws so clearly that Carlo Rubbia, the head of the UA1 group, could proclaim with full confidence: We have discovered the W particles!

During a third running period in 1983, this time with a luminosity six times greater, the Z^0 particle was discovered, along with many more W events. As a by-product, several unexplained events culled from the millions which had been recorded during the two runs were eventually identified as possible signs of the long-sought-after sixth quark (the t quark). This exciting news was announced by the UA1 group in July 1984.

Thus, in a few fruitful months, the $p\bar{p}$ collider at CERN not only discovered the weak force carriers, but might also have completed the list of quarks. Several months later the 1984 Nobel prize was awarded to Carlo Rubbia and Simon Van der Meer of CERN. The first was the driving force behind the $p\bar{p}$ collider idea, and the leading figure in the search for the Ws and Z in CERN. Van der Meer was the inventor of the 'stochastic cooling' technique (see below) which allowed the construction of the $p\bar{p}$ collider and was also in charge of the technical work of building the machine.

Let us have a closer glance at the machine and the discoveries made therein.

The CERN proton–anti-proton collider

The complex of facilities used to produce proton–anti-proton collisions in the CERN collider is schematically depicted in Fig. 10.14. The first stage in this sequence of interconnected accelerator rings is the 28-GeV Proton Synchrotron (PS). Completed in 1959 this was

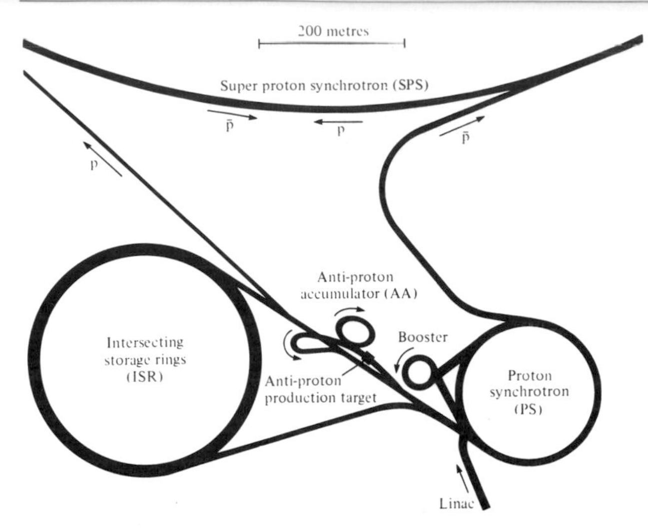

Figure 10.14. The facilities used for $p\bar{p}$ collisions at the Super Proton Synchrotron (SPS) collider at CERN. Protons are initially accelerated in the linac and booster, and then in the Proton Synchrotron (PS), from which they are transferred to the SPS ring. Anti-protons are formed by collisions of protons with a metal target, and go to the Anti-proton Accumulator for the 'cooling' process, after which they are accelerated in the PS and sent to the SPS. The intersecting storage rings to the left do not participate in that process.

the largest proton accelerator in the world (until, two years later, a 33-GeV proton synchrotron was inaugurated at Brookhaven). Now it serves as the source of protons for all the CERN devices.

In order to produce the anti-protons, protons are extracted from the PS and directed at a metal target. In the collisions some anti-protons are produced (among other particles) at about 3.5 GeV. They are collected and transferred to a ring called the Anti-proton Accumulator (AA). In this ring the important process of 'stochastic cooling' is carried out, about which a few words should be said. When the anti-protons are created, their energies and directions are quite dispersed. Prior to further acceleration they have to be stacked into dense, highly ordered bunches. One can look on the original anti-proton beam as a gas which is too 'hot' and thus reducing its random motion is referred to as 'cooling'.

The technique of 'stochastic cooling' used in the CERN collider was developed by Simon Van der Meer in 1968. It is based on two sorts of electronic devices. One is a sensing or 'pickup' device which measures the deviation of the charged particles from the desired orbit. It sends a correction signal to another device called a 'kicker' on the opposite side of the AA ring, which then applies an appropriate electric field to correct the deviation. The speed of the particles is nearly that of light, but the correction signal is trans-

Figure 10.15. Part of the UA1 detector at CERN. (Courtesy CERN.)

mitted by a wire that takes a shortcut across the chord of the ring. The signal thus reaches the kicker before the arrival of the beam, and the correction field is applied just in time. Several pairs of linked pickup and kicker devices are installed along the AA ring, in addition to the bending and focusing magnets found in any storage ring.

After the cooling process the anti-proton bunches are injected again into the PS ring (via the elongated loop to the left of the AA ring, see Fig. 10.14), and accelerated to 26 GeV. The anti-protons are then sent to the 7-kilometre-circumference SPS ring, where protons that were accelerated in the PS to the same energy are already revolving in the opposite direction. In the SPS the counter-rotating beams of particles and anti-particles are brought up to 270 GeV each.

The detectors The two huge detectors involved in the discovery of the intermediate vector bosons are called UA1 and UA2 (UA stands for underground area). Placed underground, in two sections of the SPS where collisions between the beams take place, the two detectors function independently of each other. Each of them encircles the tube in which the beams move, so that most of the particles produced in a collision (except those emerging at very small angles to the beam direction) must pass through the detector.

UA1 is 10 metres long by 5 metres wide, weighing some 2000 tonnes (see Fig. 10.15). It contains several layers of detectors in a

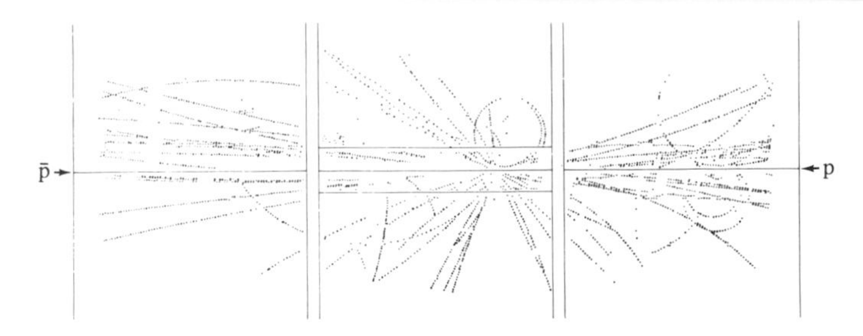

Figure 10.16. A visual display of the tracks of charged particles produced in a single $p\bar{p}$ annihilation, as recorded by the UA1 detector. (Courtesy CERN.)

concentric geometry, all triggered to operate when a $p\bar{p}$ collision occurs. The central part is a cylindrical detector of the 'drift chamber' sort, which wraps around the collision zone. In it, an array of closely spaced wires is strung in a gas at low pressure. When a charged particle passes through the chamber, it causes ionization of the gas molecules and the ions drift to a nearby wire. The wires are connected to an electronic system which records the charge distribution in the chamber. The data are analysed by a computer to reconstruct the trajectories of the charged particles emerging from the point of collision. The whole picture is stored in the computer memory and can later be displayed to give a visual record of the event (see Fig. 10.16). The accuracy in reconstructing the trajectories is remarkable, justifying the nickname 'digital bubble chamber' given to this part of the UA1 detector.

The drift chamber is surrounded by detectors (calorimeters) which measure the energy losses of charged particles passing through matter, and by scintillation counters. An 800-tonne magnet produces a magnetic field in the inner detector. The curvature of the particle path due to the magnetic field helps calculate its energy. An outer layer of drift chambers detects muons, which penetrate through all the inner layers.

The UA2 is a similar but smaller array of detectors, designed specifically to detect the intermediate bosons. It has no magnet and weighs only 200 tonnes.

Discovering the W and Z bosons When the first $p\bar{p}$ collisions made their appearance at the CERN SPS, the researchers already had detailed plans on how to search for the $W^{\pm}$ and Z^0 particles.

It was assumed that once a W particle is produced, it should decay within a very short interval (of the order of 10^{-20} seconds) to hadrons or leptons. To try to detect the W by its hadronic decay products seemed preposterous because of the myriad of other hadron-generating events in such $p\bar{p}$ collisions. It was therefore decided to look for a very special decay mode: the one in which the W turns into a single electron (or positron) and a neutrino:

$$W^+ \rightarrow e^+ + \nu_e, \quad W^- \rightarrow e^- + \bar{\nu}_e$$

The instructions for the computer were to select events in which:

(1) a considerable part of the collision energy is carried off by an isolated electron, moving in a direction transverse to the beam axis;
(2) another considerable amount of energy is carried in the opposite direction, by a neutral particle which evades the detectors (a recognition mark of the neutrino).

During the 30-day running period in November and December 1982, each of the UA1 and UA2 groups observed about 10^9 $p\bar{p}$ collisions, a million of which were recorded. A computer program was then operated in order to discard any event that might have originated in a process other than the decay of the W. In the first stage, 140 000 events of the UA1 harvest were culled, out of which only 2125 stood the criteria of another selection. After additional careful examinations, the sample was reduced to 39 events, and each of these was analysed and interpreted separately. At the end of this painstaking process five events were found which could only be interpreted as 'W-events': four in which a W^- decayed into an electron and an anti-neutrino and one in which a W^+ turned into a positron and a neutrino.

A similar analysis of the UA2 data yielded four additional W decays. The mass of the W was estimated to be about 81 GeV, in excellent agreement with the electroweak theory prediction.

The search after the Z^0 was based on the decay modes: $Z^0 \rightarrow e^+ + e^-$, $Z^0 \rightarrow \mu^+ + \mu^-$. The events looked for were those in which high-energy electron and positron (or positive and negative muons) moved in opposite directions, perpendicular to the beam axis. The production rate of the Z^0 in $p\bar{p}$ collisions was predicted to be lower than that of the Ws and it was no surprise when such events were discovered during the first run of the SPS collider. In the subsequent run, however, the luminosity was six times higher and the two groups began to record clear Z^0 events. Up to mid-1983 several such events were gathered and analysed, and the discovery of the Z^0 was then officially announced. The mass estimated from the data was quite close to the theoretical value of 94 GeV. In the same run many more W particles were detected, and additional support to the assumption that these were really the weak force carriers was obtained. It was anticipated that because of parity violation in weak interactions, there would be a strong backward–forward asymmetry in the directions of the charged leptons produced in the W decay: the electrons (or μ^-) would be preferentially emitted in the direction of the proton beam and the positrons (or μ^+) in the direction of the anti-protons. When the harvest of W events amounted to about 30, this asymmetry was very apparent.

The t quark at last? The thorough scanning of the two million interactions recorded between November 1982 and July 1983 revealed several 'anomalous events'. In some of them a charged lepton (electron or muon) was recorded together with the missing energy characteristic of a neutrino, but these were accompanied by two narrow jets of particles. These events were interpreted as the decay of a W into a bound system of quark and anti-quark, which in turn decayed to the observed final products. Mass calculations showed that one of the quarks was the already known b quark, and that the mass of the other was in the range of 30–50 GeV, much greater than that of any known quark. When several such events were assembled, all showing the same features, it became plausible that the discovery of the long-sought t quark had been made. A typical such event was interpreted in the following way:

$$
\begin{aligned}
W^+ &\rightarrow t + \bar{b} \\
& \qquad\quad \hookrightarrow \text{jet} \\
& \quad \hookrightarrow e^+ + \nu_e + b \\
& \qquad\qquad\qquad\quad \hookrightarrow \text{jet}
\end{aligned}
$$

or with a muon instead of the positron. The mass of the t quark was estimated to be around 40 GeV.

10.9 The standard model and beyond

After the discovery of the particles W^+, W^- and Z^0 and the evidence for the existence of the t quark, the so-called 'Standard model' of particle physics appears well established and vindicated. According to this model the elementary particles are leptons and quarks. Both have spin $\frac{1}{2}$ (and are therefore fermions) and are apparently pointlike. The basic difference between them is that the quarks are affected by the strong force (known also as the 'chromodynamic force' described by 'quantum chromodynamics' or QCD), which does not influence the leptons. All known quarks and leptons can be arranged in families or 'generations' of increasing mass, as shown in Table 10.4. Each generation consists of a pair of leptons (one charged and one neutral) and a pair of quarks (carrying charges of

Table 10.4. *Generations of quarks and leptons*

	1		2		3		
Quarks	d	u	s	c	b	t	?
Leptons	e	ν_e	μ	ν_μ	τ	ν_τ(?)	⟶

$-\frac{1}{3}$ and $+\frac{2}{3}$). For each pair there is a corresponding pair of antiparticles. The first generation includes the quarks u, d and the leptons e, ν_e. Since u and d are the building blocks of the proton and neutron, this generation actually includes all the elementary constituents of ordinary matter. The other two generations were discovered and studied in high-energy experiments or in cosmic rays. Additional generations might be discovered when, and if, higher energies become accessible.

The putative carrier of gravitation has a spin of two units.

According to the standard model the basic forces, or interactions, between the elementary particles are carried by spin-one bosons. The electromagnetic and weak forces are well described by the electroweak theory of Weinberg, Salam and Glashow, which reveals the common origin of those forces. The force-carrying particles in this theory are the massive particles W^+, W^-, Z^0 and the massless photon. The strong force between the quarks is described by the QCD theory which is based on eight electrically neutral carrier particles – the gluons.

An important concept which should be mentioned here is the gauge theory. This name indicates a type of quantum field theory in which the field equations do not change when we perform some operations on a family of particles everywhere in space. (Actually it is required that the equations remain invariant even if we perform those operations independently at each point in space and time.) We cannot go further into this subject, except for mentioning that the theories of all the basic forces can be formulated, in one way or the other, as gauge theories, and this may lead us eventually to a coherent unified theory.

Higgs bosons and magnetic monopoles

Yet the standard model cannot be considered complete either from the theoretical or from the experimental point of view. For example, the theoretical proof of the quarks' confinement using QCD, so needed to understand why free quarks have not been seen, has not yet been achieved. In addition, this theory does not provide an explanation for the masses of the elementary particles, or of the relations between the various 'generations' of quarks and leptons.

In the experimental domain, numerous predictions of the standard model concerning the occurrence of certain reactions, estimated cross-sections, distribution of products, etc., as well as the energy spectrum of two-quark states, are yet to be confirmed. One of those predictions pertains to particles known as 'Higgs bosons', named after Peter Higgs of the University of Edinburgh. Higgs bosons appear in the electroweak theory as massive particles which couple to the carriers of the weak force, providing them with mass, but not to the photon, which is thus left massless. Their masses are not known but according to the theory they may manifest themselves on rare occasions as free particles.

Attempts to go beyond the standard model are continually made. The theories of grand unification or grand unified theories (GUTs) which seek to unify the electroweak and strong interactions in one coherent structure are examples of such attempts. These theories predict various new phenomena which can be looked for experimentally. The instability of the proton is one example (see section 7.6). Other predictions speak of new and exotic particles such as magnetic monopoles: particles carrying a 'magnetic charge' instead of, or in addition to, the electric charge.

In our known physical world there is an obvious asymmetry between electricity and magnetism since there are electric charges but no magnetic charge. (All the known magnetic phenomena appear to be caused by moving electric charges.) It was Paul Dirac who, back in 1931, pointed out that the existence of 'magnetic charges' would require some modifications of Maxwell's equations, but would not contradict any fundamental physical law, and would provide a nice explanation of the fact that electric charges always appear as integer multiples of a basic unit. A hypothetical particle carrying the magnetic charge was named the 'magnetic monopole' (a conventional magnet is always a dipole, having two magnetic poles). The grand unification theories have revived the interest in Dirac's monopoles. According to these theories massive magnetic monopoles might have been produced in the first seconds following the big bang, which created our universe. Some of these may have survived and can perhaps be trapped and detected. Several cosmic ray events could be interpreted as the signature of magnetic monopoles. However, the rarity of these occurrences has so far left the magnetic monopole in the position of a hypothetical particle which has not yet been discovered.

Some theories which go beyond the standard model try to describe the quarks and leptons as composite particles built up of even more fundamental constituents, as will be discussed in section 10.11.

10.10 Future accelerators

In order to check the various theories and speculations which take us beyond the standard model, as well as the predictions of the standard model itself, new particle accelerators are needed. It is not just a matter of higher energies, but of higher luminosity (particle flux per unit area in the cross-section of the beam) and better sensitivity of the detectors as well. To illustrate this point we shall mention that certain theoretical predictions can be checked by investigating the decay of the Z^0. However, for this purpose an abundant source of Z^0 particles is needed, much more than the 540-GeV CERN $p\bar{p}$ storage ring can provide (its total was less than two dozen Z^0 bosons in a whole year). The new 100-GeV e^+e^- colliders at SLAC and at CERN are supposed to give more than that in just *one day* and allow a thorough examination of the Z^0 decay products.

What are the accelerators that will dominate particle research in the next few years? Most of them will be colliders of various sorts. One representative of this new generation of colliders is the e^+e^- LEP storage ring under construction at CERN. This gigantic ring (27 kilometres in circumference) will cost several hundred million dollars to build and will commence operation in 1989, at a total energy of 100 GeV (50 GeV per beam). Apart from the size, beam energy and some technical details, the LEP is essentially very similar to previous storage rings, such as SPEAR, which was described in section 8.4. There are two reasons for the larger size of this ring. First, the magnetic field needed to retain a particle in a circular path of a given radius grows with the energy of the particle, and is smaller when the radius is larger. At high energies it becomes cheaper to construct a bigger ring than to use very high magnetic fields. The second reason is related to the phenomenon of 'synchrotron radiation'.

We have mentioned (section 4.2) that at relativistic velocities, the rate at which the kinetic energy of the charged particle is dissipated by the emission of synchrotron radiation is proportional to E^4/M^4R^2 where E and M are the energy and mass of the particle, and R is the radius of its orbit. Since the fourth power of M appears in the denominator, the radiated power is very much greater (by a factor of 10^{13} actually) for the light electron than for the heavier proton. These energy losses must be continually replenished and at high energies this may be an unattainable accomplishment. Increasing the ring's radius serves to mitigate the dissipation (recall that the power radiated is proportional to $1/R^2$).

Because of the synchrotron radiation, some physicists think that electron–positron storage rings of higher energies than that of the

Figure 10.17. An aerial photograph of SLAC. The linear accelerator passes under the highway, SPEAR and PEP are in the centre and the dashed line marks the path of the new linear collider (SLC). (Courtesy SLAC.)

LEP would be impractical, and that future e^+e^- colliders will employ two opposite linacs. These linacs will fire electron and positron bunches into a tube between them, and the bunches will collide just once.

This novel idea may be tested in Novosibirsk, USSR, where a linear e^+e^- collider of 300 GeV total energy is planned. In the meantime SLAC is constructing a machine which, although called the SLAC Linear Collider (SLC), may well be referred to as a 'semi-linear' collider. It is planned to achieve 100 GeV several years before the LEP, and at much lower cost; giving the Americans once more the precedence in accelerator technology. The idea, proposed by B. Richter, was to utilize the existing electron linac, upgraded to 50 GeV, to propel successive bunches of electrons and positrons in the same direction by the opposite phases of the alternating voltage used to accelerate particles along the tube. At the outlet of the 2-mile linac the e^- and e^+ bunches will be separated and driven in opposite directions around a ring-like course, to collide head on. Since the particles traverse the ring only once, the energy loss through synchrotron radiation will be much smaller than that in the storage rings. The construction of SLC began in 1984 and is scheduled to be completed by 1986 (see Fig. 10.17).

It was mentioned that synchrotron radiation is much smaller for protons than for electrons. Thus the Fermilab Tevatron can retain 1000-GeV protons in a ring less than one-fourth the size of the LEP. By 1986 this American machine is going to operate as a $p\bar{p}$ storage ring of 2000 GeV total energy. It will exhibit not only higher energy than the $p\bar{p}$ storage ring of CERN, but higher luminosity as well (about 10 times larger). Can high-energy physics be pushed even higher, say, to the 10–20 TeV range? Physicists do not see any technical difficulty in principle, apart from the financial one. The High Energy Physics Advisory Panel in the USA has recommended the building of a Superconducting Supercollider (SSC) in which beams of 20-TeV protons, or protons and anti-protons, would collide head on. It will cost more than 3×10^9 dollars and take more than a decade to construct. Since this immense machine will require a vast area of open land, it has already been nicknamed 'The Desertron'. By the turn of the century, a similar 10–20-TeV proton synchrotron may lodge in the LEP tunnel at CERN. Bigger accelerators would be so expensive to build and maintain that a worldwide international effort might be inevitable. A world accelerator is a seriously discussed topic of international conferences and committees. The smooth operation of CERN over so many years provides convincing evidence that this can be done.

International cooperation has become popular at other laboratories as well. At DESY in Hamburg, groups from several countries have been working in harmony for some years. When planning the new HERA project at DESY, the German authorities called for the participation of other countries, allowing the foreign investments to be spent in the countries providing them. Various parts of HERA are now being built in Germany, France, Italy, Holland, Canada, Israel and other countries. In this machine (to be completed by 1990) 30-GeV electrons will collide with 820-GeV protons, in a tunnel of 6.3 kilometres circumference, to provide a centre-of-mass energy of 314 GeV. Many physicists consider it to be a very promising project. For one thing, the collisions will probe the structure of the proton to a separation of 10^{-17} centimetres and check whether quarks look pointlike at this close distance. Thus HERA may be a landmark not only in international cooperation but in high-energy physics as well.

Hera was the queen of gods in the Greek mythology, but it is also the initials of Hadron–Electron Ring Accelerator.

10.11 Open questions

What is the nature of the discoveries that the new accelerators might expose? Will they find just more generations of quarks and leptons, or will entirely new phenomena emerge? No one can tell at present. Other unsolved puzzles may be resolved with some 'hints' from experimentation: why do quarks and leptons have the masses they have, and how many 'generations' of quark and lepton pairs are there? These questions are related to the deeper problem of possible inner structure within the quarks and leptons. Until now, the search for building blocks of matter has led us to more and more 'elementary' entities – from the molecule to the atom, to the nucleus and electrons, to the nucleons, and eventually to the quarks. Are we over with this 'onion peeling' process, or should we now look for the constituents of the quarks and leptons themselves? According to our present experimental data, quarks and leptons really look elementary. On the other hand, the natural explanation of the higher 'generations' of quarks and leptons is that they correspond to excited states of the first generation, and our experience suggests that an excited system must be composite.

Indeed, several models have been suggested in which the quarks and leptons are regarded as being made of more fundamental particles. One such model was proposed in 1979 by Haim Harari of the Weizmann Institute of Science in Rehovot. The model assumes that both quarks and leptons are constructed of just two types of 'ultimate' particles which Harari called rishons. (*Rishon* in Hebrew means first or primary.) The two rishons are designated T and V for *Tohu* and *Vohu* ('formless' and 'void' in Hebrew. This is the description of the universe in its initial state, according to the first chapter of Genesis). T carries an electric charge of $+\frac{1}{3}$, while V is neutral. There are also anti-rishons $\bar{\mathrm{T}}$ and $\bar{\mathrm{V}}$, possessing charges of $-\frac{1}{3}$ and 0 respectively.

A similar model was proposed independently by M. A. Shupe of the University of Illinois at Urbana-Champaign. The papers of Harari and Shupe appeared side by side in the same issue of *Physics Letters*.

In Harari's model any three rishons and any three anti-rishons may be combined together, but rishons and anti-rishons do not intermingle. This rule allows 8 combinations which correspond to the 8 quarks and leptons of the first generation and their anti-particles. The combination TTT is a positron and $\bar{\mathrm{T}}\bar{\mathrm{T}}\bar{\mathrm{T}}$ is an electron, while VVV and $\bar{\mathrm{V}}\bar{\mathrm{V}}\bar{\mathrm{V}}$ correspond to the neutrino and anti-neutrino. The other allowed combinations yield the fractionally charged quarks: TTV = u, TVV = $\bar{\mathrm{d}}$, $\bar{\mathrm{T}}\bar{\mathrm{T}}\bar{\mathrm{V}}$ = $\bar{\mathrm{u}}$ and $\bar{\mathrm{T}}\bar{\mathrm{V}}\bar{\mathrm{V}}$ = d.

By ascribing 'colours' (the 'charges' of the chromodynamic force) to the rishons, the model can nicely account for the fact that leptons do not carry colours but quarks do. So far, however, this model has failed to explain the occurrence of the higher generations of quarks and leptons; nor can it provide a satisfactory explanation for the masses within the first generation and for several other parameters that are not given by either QCD or the electroweak theory. This flaw is common to most of the models which attempt to elucidate the structure of quarks and leptons, and at present it seems that the efforts in this direction have reached a dead end. It is possible that behind these problems there lies a completely new interaction.

Another direction might be the unification of the known basic forces, especially with the prospect of a 'complete' unification, including gravity. Some physicists believe that when this project is completed, we shall have before us that final physical theory for which generations of scientists have striven. This view was expressed, for example, by Stephen Hawking of Cambridge University, who believes that the formulation of such a theory might be in sight, and that it will mark 'the end of theoretical physics'. Other physicists, however, regard this opinion as over-optimistic (or over-pessimistic, depending upon one's philosophy) and refuse to believe that we are going to fathom *all* the secrets of nature in the near future. The merger with gravity would require understanding its behaviour at the atomic level. It has recently been shown that for very short distances one would have to modify Einstein's general relativity.

Some researchers point at cosmology as a domain full of phenomena that might cast new light on fundamental issues of particle physics. We have already mentioned that theories on the amounts of matter and anti-matter in the early universe are related to the problem of conservation or non-conservation of the baryon number. Another exciting prospect is that astronomical observations might set a limit on the total number of quark and lepton 'generations'. There are in fact cosmological reasons to think that this number is not larger than four. At the same time our new understanding of particle interactions has brought about the emergence of a new cosmological model – the 'inflationary universe' – explaining many previously mysterious facts.

With the unresolved mysteries, with the fascinating prospects of cosmological applications, and with those huge manmade accelerators under construction or on drawing boards, our picture of the world might well evolve greatly in another 5, 10 or 20 years. The perspectives may be totally new and different. The answers, however, are all waiting in nature, and all we should do is go on with the search.

Appendix 1: Properties of semi-stable particles

Class	Particle	Mass (MeV)	Lifetime (seconds)	Charge	Spin	Isospin	Parity
Leptons	e^-	0.511	Stable	−1	$\frac{1}{2}$		
	ν_e	0?	Stable	0	$\frac{1}{2}$		
	μ^-	105.66	2.2×10^{-6}	−1	$\frac{1}{2}$		
	ν_μ	0?	Stable	0	$\frac{1}{2}$		
	τ^-	1784	3.4×10^{-13}	−1	$\frac{1}{2}$		
	ν_τ	0?	Stable?	0	$\frac{1}{2}$		
Mesons							
$S=0$	π^+, π^-	139.57	2.6×10^{-8}	±1	0	1	−1
	π^0	134.96	0.83×10^{-16}	0	0	1	−1
	η^0	548.8	3.7×10^{-19}	0	0	0	−1
$S \neq 0$	K^+, K^-	493.67	1.2×10^{-8}	±1	0	$\frac{1}{2}$	−1
	$K^0, \bar{K}^0$	497.67	8.9×10^{-11} (K^0_S) 5.18×10^{-8} (K^0_L)	0	0	$\frac{1}{2}$	−1
$C \neq 0$	D^+, D^-	1869.4	9.2×10^{-13}	±1	0	$\frac{1}{2}$	−1
	$D^0, \bar{D}^0$	1864.7	4.4×10^{-13}	0	0	$\frac{1}{2}$	−1
$S \neq 0, C \neq 0$	F^+, F^-	1971	$\sim 2 \times 10^{-13}$	±1	0	0	−1
$B \neq 0$	B^+, B^-	5271	$\sim 1.5 \times 10^{-12}$	±1	0	$\frac{1}{2}$	−1
	$B^0, \bar{B}^0$	5274	$\sim 1.5 \times 10^{-12}$	0	0	$\frac{1}{2}$	−1
Baryons							
$S=0$	p	938.28	Stable	1	$\frac{1}{2}$	$\frac{1}{2}$	+1
	n	939.57	920	0	$\frac{1}{2}$	$\frac{1}{2}$	+1
$S=-1$	Λ^0	1115.6	2.63×10^{-10}	0	$\frac{1}{2}$	0	+1
	Σ^+	1189.4	0.8×10^{-10}	+1	$\frac{1}{2}$	1	+1
	Σ^0	1192.5	5.8×10^{-20}	0	$\frac{1}{2}$	1	+1
	Σ^-	1197.3	1.5×10^{-10}	−1	$\frac{1}{2}$	1	+1
$S=-2$	Ξ^0	1314.9	2.9×10^{-10}	0	$\frac{1}{2}$	$\frac{1}{2}$	+1
	Ξ^-	1321.3	1.64×10^{-10}	−1	$\frac{1}{2}$	$\frac{1}{2}$	+1
$S=-3$	Ω^-	1672.4	0.82×10^{-10}	−1	$\frac{3}{2}$	0	+1
$C=1$	Λ_c^+	2282	2.3×10^{-13}	+1	$\frac{1}{2}$	0	+1
Intermediate bosons	γ	0	Stable	0	1		
	$W^\pm$	~80800	?	±1	1		
	Z^0	~92900	?	0	1		

Appendix 2: The Greek alphabet

Α	α	Alpha	(ä)
Β	β	Beta	(b)
Γ	γ	Gamma	(g)
Δ	δ	Delta	(d)
Ε	ε	Epsilon	(e)
Ζ	ζ	Zeta	(z)
Η	η	Eta	(ä)
Θ	θ	Theta	(th)
Ι	ι	Iota	(ë)
Κ	κ	Kappa	(k)
Λ	λ	Lambda	(l)
Μ	μ	Mu	(m)
Ν	ν	Nu	(n)
Ξ	ξ	Xi	(ks)
Ο	ο	Omicron	(o)
Π	π	Pi	(p)
Ρ	ρ	Rho	(r)
Σ	σ ς	Sigma	(s)
Τ	τ	Tau	(t)
Υ	υ	Upsilon	(öö)
Φ	φ	Phi	(f)
Χ	χ	Chi	(kh)
Ψ	ψ	Psi	(ps)
Ω	ω	Omega	(ö)

Name index

Subject index